CW01149921

FABER'S NEW ANATOMICAL ATLAS

other titles of interest published by Faber

THE NURSE'S DICTIONARY

FABER POCKET MEDICAL DICTIONARY

AN A TO Z OF GYNAECOLOGY
by Mary Anderson

THE ANATOMY AND PHYSIOLOGY OF OBSTETRICS
by C. W. F. Burnett, revised by Mary Anderson

MIDWIFERY QUESTIONS ANSWERED
edited by Valerie Smith

SPECIAL TESTS AND THEIR MEANINGS
By D. M. D. Evans

HUMAN BEHAVIOUR IN ILLNESS
by Lynn Gillis

PRINCIPAL DRUGS: AN ALPHABETICAL GUIDE
by S. J. Hopkins

ANATOMY AND PHYSIOLOGY FOR NURSES
by Evelyn Pearce

MEDICAL CALCULATIONS FOR NURSES
by G. W. Watchorn

FABER'S NEW ANATOMICAL ATLAS

Anne Roberts, MB, BS, MRCP
and
Audrey Besterman, FMAA

ff
faber and faber
LONDON · BOSTON

First published in 1988
by Faber and Faber Limited
3 Queen Square London WC1N 3AU

Photoset by Parker Typesetting Service Leicester
Printed in Great Britain by W S Cowell Ltd Ipswich
All rights reserved

Text © Anne Roberts, 1988
Illustrations © Audrey Besterman, 1988

*This book is sold subject to the condition that it shall not,
by way of trade or otherwise, be lent, resold, hired out
or otherwise circulated without the publisher's prior consent
in any form of binding or cover other than that in which
it is published and without a similar condition including this
condition being imposed on the subsequent purchaser.*

British Library Cataloguing in Publication Data

Roberts, Anne
The new Faber anatomical atlas.
1. Anatomy, Human
I. Title II. Besterman, Audrey
611   QM23.2
ISBN 0-571-13839-X

Contents

PLATES

1. (*a*) The meaning of the terms used in anatomy.
   (*b*) Structures and their functions.
2. The skeleton.
3. The skeleton of the head and trunk from behind.
4. The bones of (*a*) the hand and (*b*) the foot.
5. (*a*) The lacrimal apparatus.
   (*b*) A synovial joint (diagrammatic).
6. (*a*) The names of movements.
   (*b*) Types of joint.
   (*c*) Particular joints.
7. Superficial muscles.
8. Muscles from behind (superficial ones on the left, deeper ones on the right). It also shows the bones of the right forearm pronated.
9. (*a*) The position of the paranasal air sinuses.
   (*b*) The right halves of the laryngeal cartilages.
   (*c*) The right half of the larynx.
10. (*a*) The trachea, bronchi and lungs (anterior view, left lung cut open).

- (b) The divisions of the mediastinum (diagrammatic).
- (c) A pulmonary unit (where gas exchange takes place).

11 (a) Deciduous (milk) teeth, showing usual times of eruption.
- (b) Permanent teeth, showing usual times of eruption.
- (c) A lower incisor tooth (sagittal section).
- (d) The salivary glands.
- (e) The mouth cavity from in front.

12 The thorax and abdomen from the front: superficial structures.

13 The thorax and abdomen: deeper structures.

14 The thorax and abdomen: structures on the posterior wall and pelvic contents (female).

15 (a) The head and neck (sagittal section).
- (b) The regions of the abdomen.
- (c) Transverse section of the small intestine to show the structure.
- (d) The stomach, duodenum and biliary tree.
- (e) Sagittal section of the abdominal cavity to show the peritoneum; the lesser sac is cross-hatched, the greater sac is plain.

16 (a) The right kidney, with its suprarenal gland, shown complete; the left kidney is sectioned.
- (b) A nephron (diagrammatic).

17 (a) The male urogenital system.
- (b) The male perineum; the penis is sectioned to show its structure.

18 The female urogenital system:
- (a) Sagittal section of the pelvis.
- (b) The right half of the uterus and vagina.
- (c) The perineal triangles.
- (d) The vulva.

CONTENTS

19 Some endocrine organs and their hormones (not to scale).
20 (a) A neuron.
    (b) Detail of a synapse.
    (c) The left side of the brain (semidiagrammatic).
21 Sagittal section of the central nervous system showing the circulation of the CSF (semidiagrammatic).
22 The central nervous system. Principal motor and sensory pathways (diagrammatic).
23 The distribution of some of the branches of (a) the trigeminal nerve and (b) the facial nerve. (c) The nerve supply of the tongue.
24 (a) The inside of the base of the skull showing foramina (left side) and structures passing through them (right side).
    (b) The course and distribution of the vagus nerves (from behind, semidiagrammatic).
    (c) The cavernous sinus (coronal section).
    (d) The central connections of the vestibulocochlear nerve (diagrammatic).
25 (a) Plan of the brachial plexus.
    (b) Cutaneous nerve supply of the right upper limb and course of the main nerves.
26 (a) Plan of the lumbosacral plexus.
    (b) Cutaneous nerve supply of the right lower limb and course of the main nerves.
27 The functions of the autonomic nervous system.
28 The eye:
    (a) Sagittal section of the left eye.
    (b) The optic pathway.
    (c) The extrinsic eye muscles and their nerves.

29 The ear:
- (*a*) Coronal section showing external, middle and inner ear (seen from the front).
- (*b*) Right tympanic membrane, as seen through an auriscope.
- (*c*) The right inner ear.
- (*d*) Section through the cochlea.

30 (*a*) The circulation of the blood through the heart and great vessels; red indicates oxygenated blood, blue denotes deoxygenated blood.
- (*b*) The coronary arteries.
- (*c*) Electrical conduction in the heart.

31 (*a*) Diagram of the circulation of the blood (colours as in Fig. 30).
- (*b*) Blood vessels.

32 (*a*) The heart valves, viewed from above (atria partially removed).
  LA, RA: left atrium, right atrium;
  LV, RV: left ventricle, right ventricle.
- (*b*) The porta hepatis.

33 Principal arteries.

34 The arterial supply to the abdominal and pelvic viscera.
- (*a*) Coeliac artery.
- (*b*) Superior mesenteric artery.
- (*c*) Inferior mesenteric artery.
- (*d*) Internal iliac artery.

35 The blood supply to the central nervous system:
- (*a*) Sources of arterial blood to the brain (diagrammatic).
- (*b*) The brainstem from below showing arteries and cranial nerves.
- (*c*) The spinal cord and sympathetic chains (blood supply shown on the left).

CONTENTS

36 Principal veins.

37 (a) Coronal section through the vault of the skull.
    (b) The dura mater and venous sinuses (part of the skull and the whole of the brain removed).

38 The hepatic portal system; important portosystemic anastomoses are ringed.

39 (a) Detail of a single lymph node.
    (b) Principal nodes and lymphatic drainage of the head and neck.
    (c) Principal nodes and drainage of the trunk and limbs.

40 (a) Section through the skin (much enlarged).
    (b) The lactating breast: structure shown in the right breast, lymphatic drainage from the left.

TEXT

The Skeleton	43
The Muscles	48
The Respiratory System	49
The Digestive System	54
The Urinary System	69
The Male Reproductive System	73
The Female Reproductive System	76
The Endocrine System	80
The Nervous System	84
The Special Senses	105
The Cardiovascular or Circulatory System	110
The Lymphatic System	119
The Skin and its Appendages	122

Preface

This book is designed to present the basic facts of human anatomy in a clear, concise form. Both text and illustrations are new, but the plan – clear figures with a brief explanatory text – follows that of the very popular *Faber's Anatomical Atlas*. The full figure position drawn for the main body systems has been chosen to give a three-dimensional effect and an understanding of structures that run from the front to the back of a limb – for example, the femoral artery.

The book should be useful to nurses, physiotherapists, first-aiders and others requiring essential anatomy in a simple form. It will be of interest to artists and students of human biology, and should also prove a refreshingly uncluttered revision aid for medical students. Finally, we should especially like it to be read by general readers, who are often surprisingly ignorant of those intricate and wonderful machines they use every day – their bodies.

AR
AB
1988

PLATES   I

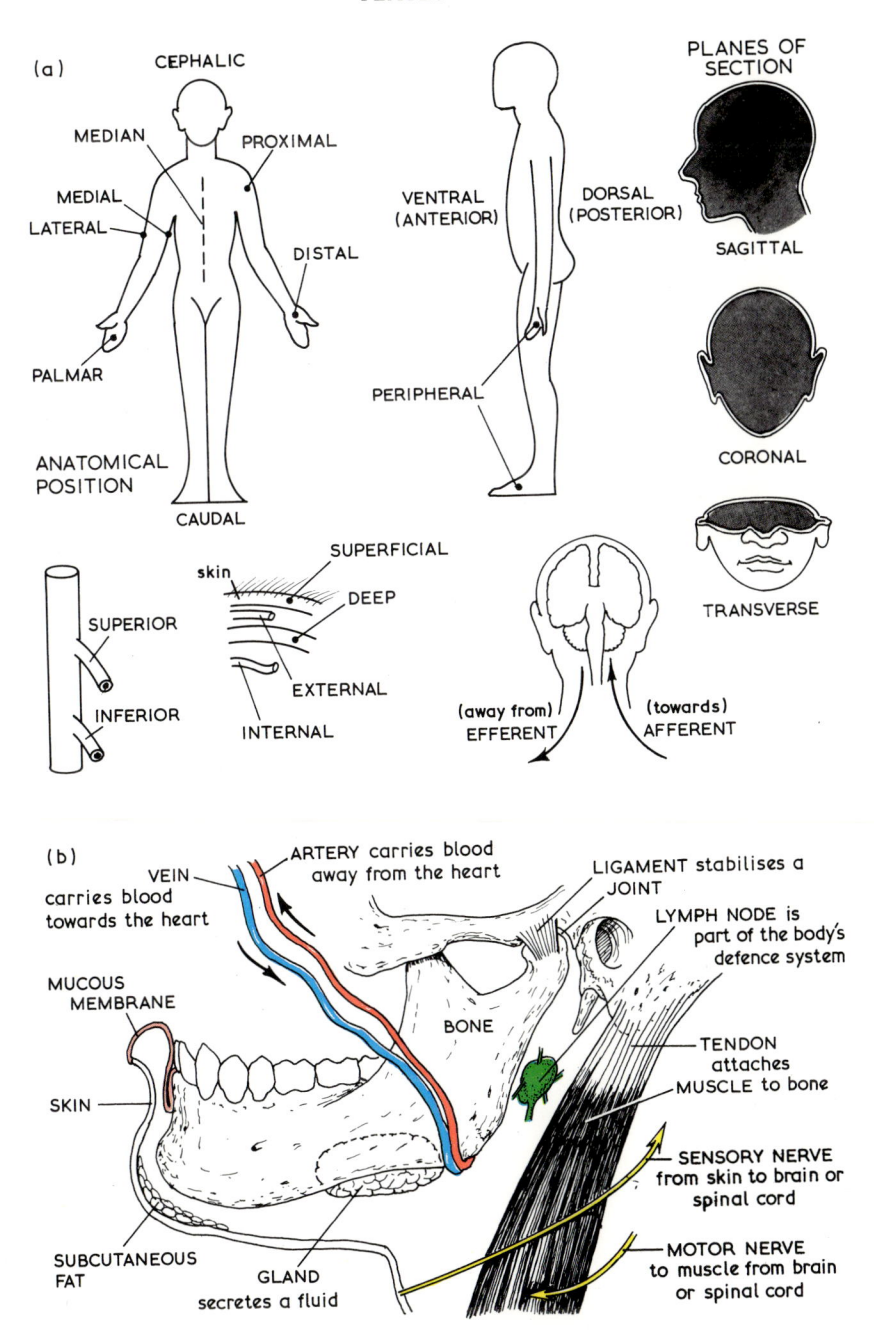

1 (a) The meaning of the terms used in anatomy.
  (b) Structures and their functions.

2 The skeleton.

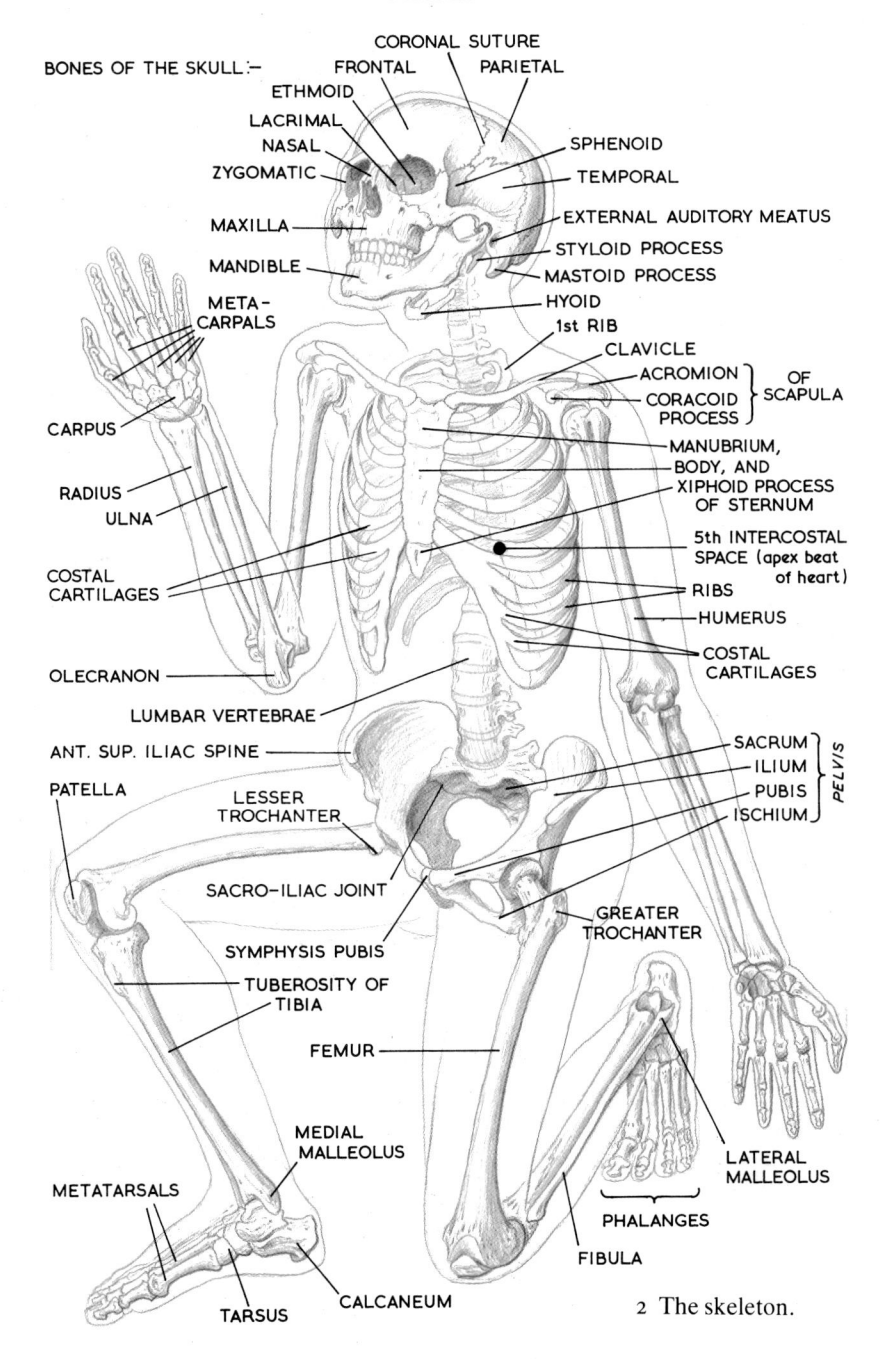

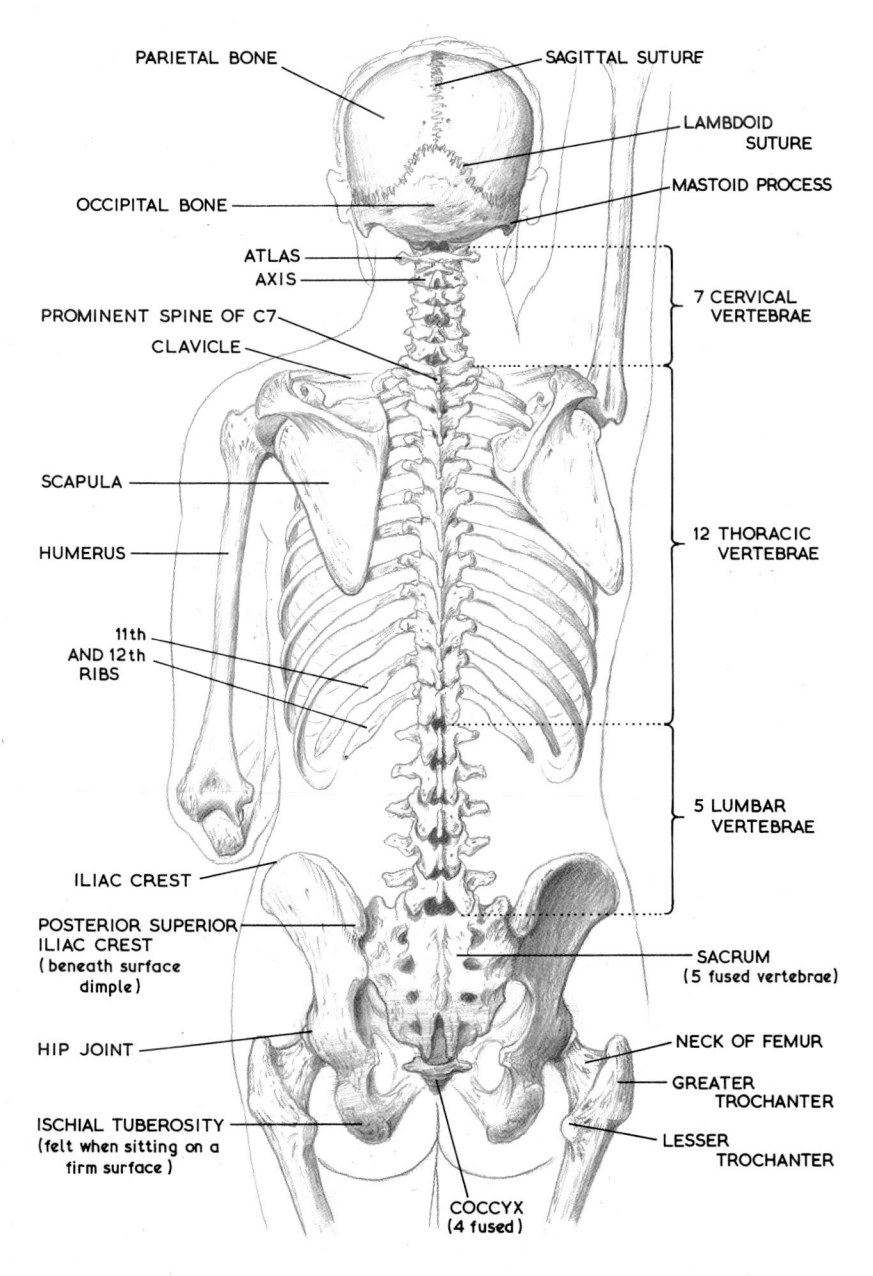

3 The skeleton of the head and trunk from behind.

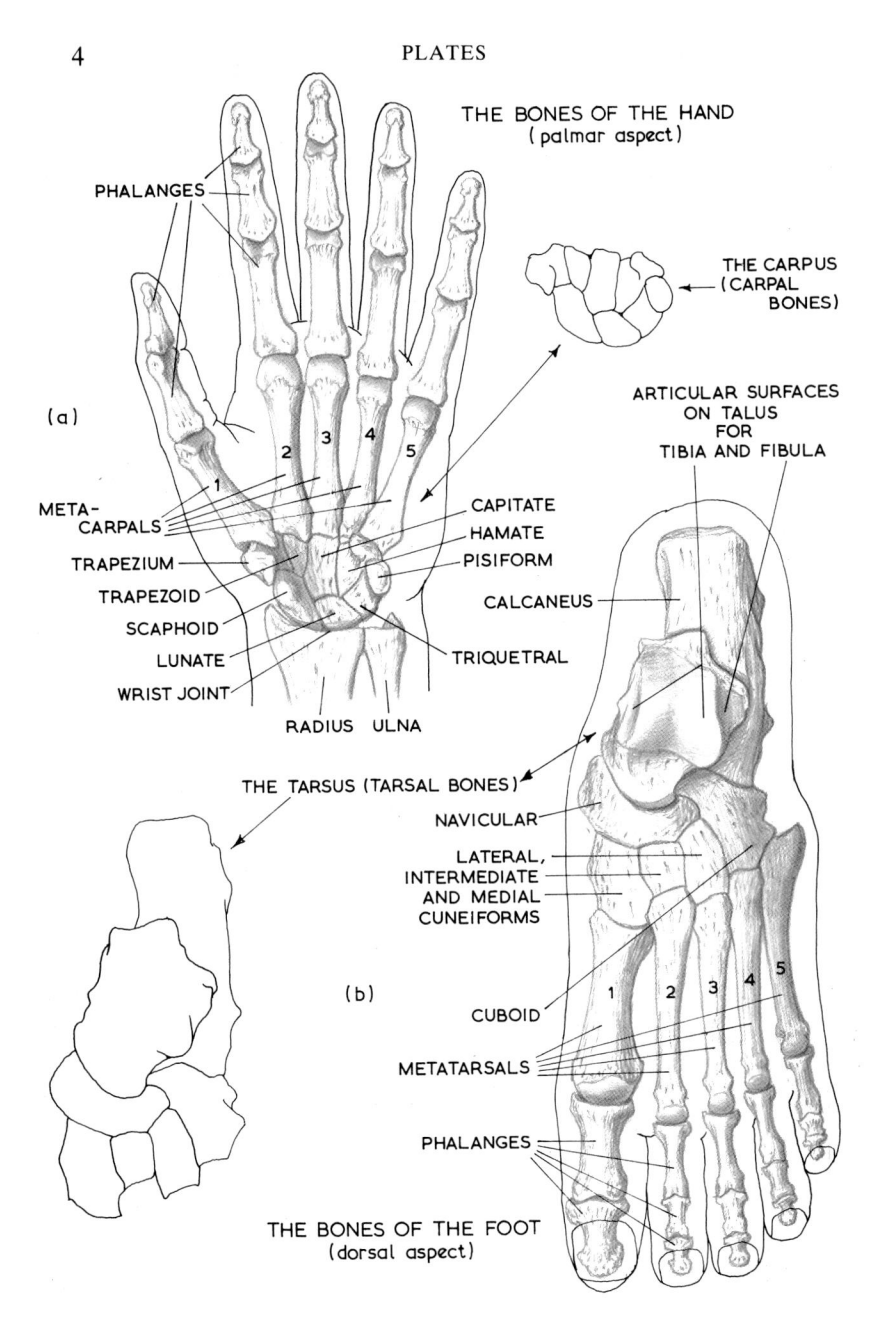

4 The bones of (*a*) the hand and (*b*) the foot.

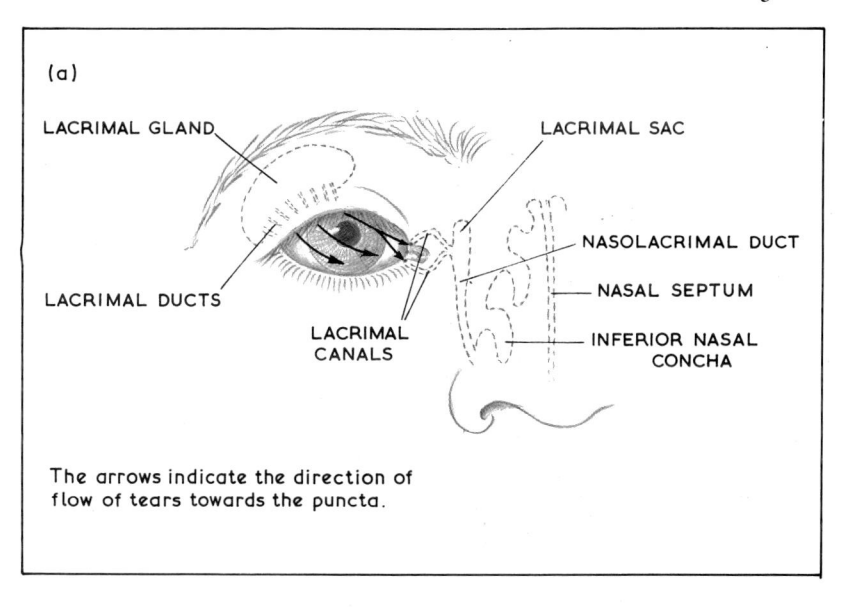

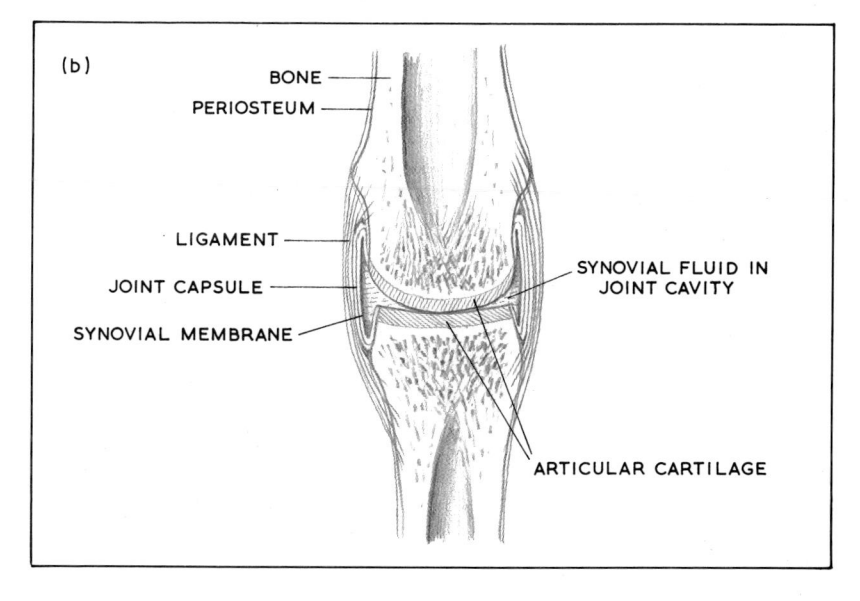

5 (a) The lacrimal apparatus.
  (b) A synovial joint (diagrammatic).

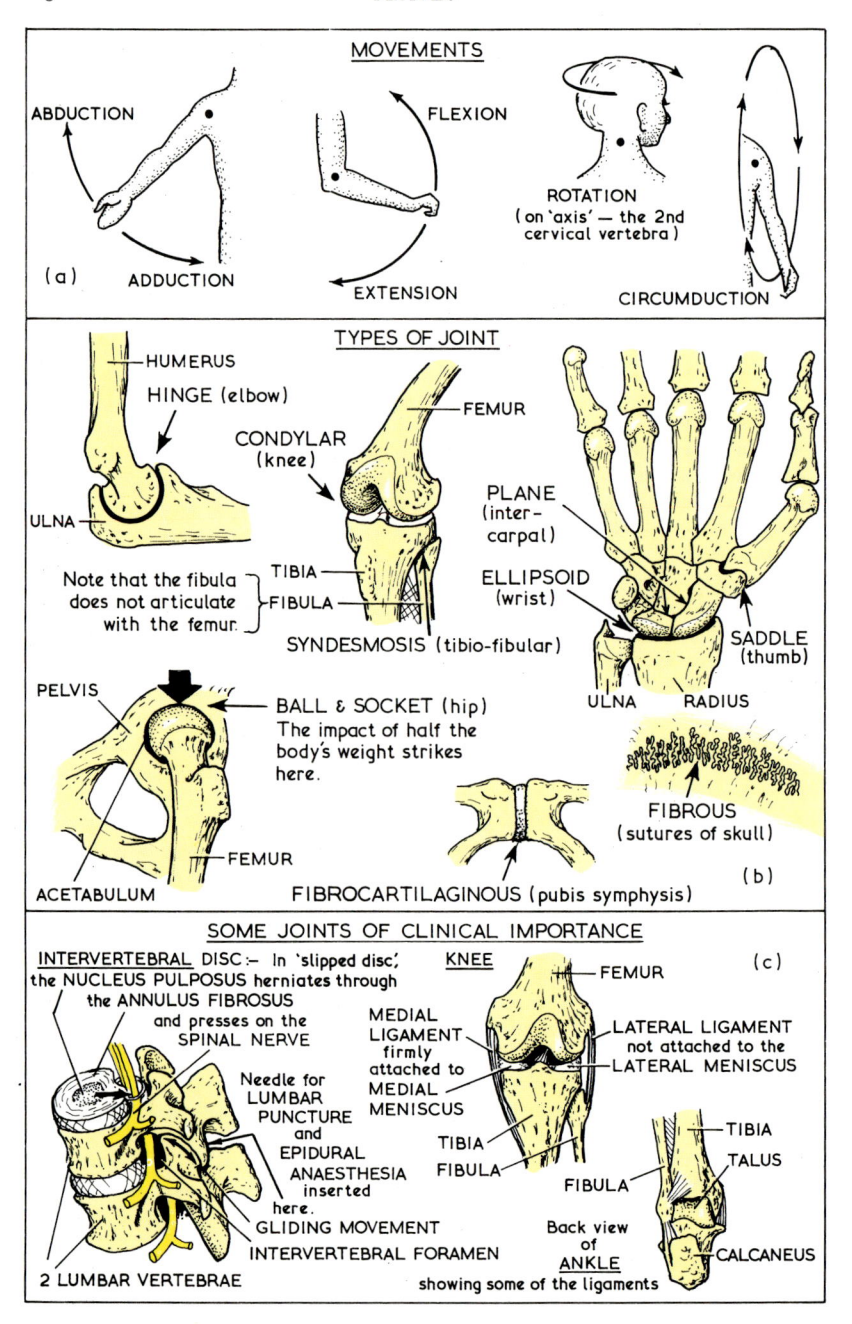

6 (a) The names of movements.
 (b) Types of joint.
 (c) Particular joints.

PLATES 7

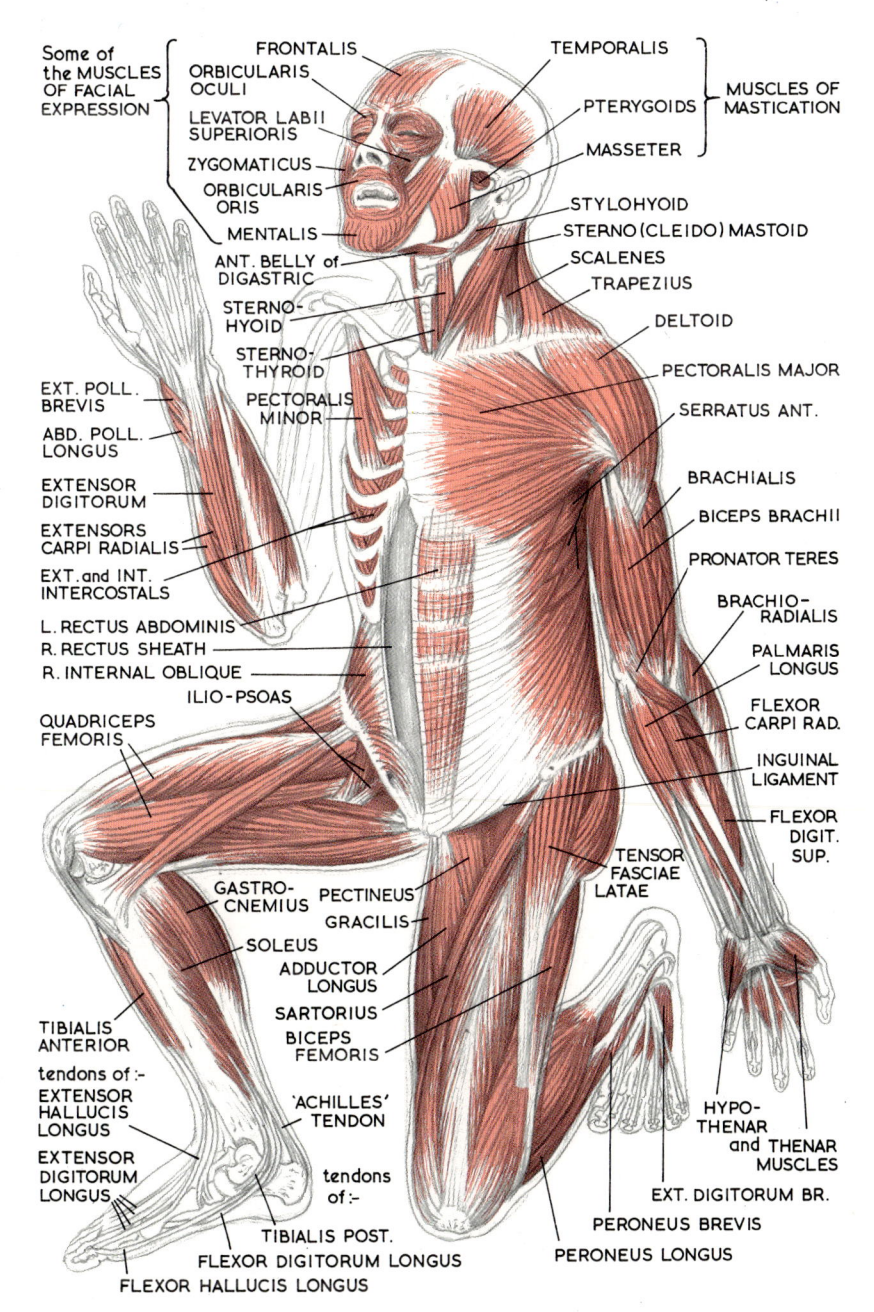

7 Superficial muscles.

8   PLATES

SUPERFICIAL MUSCLES   **DEEPER MUSCLES**

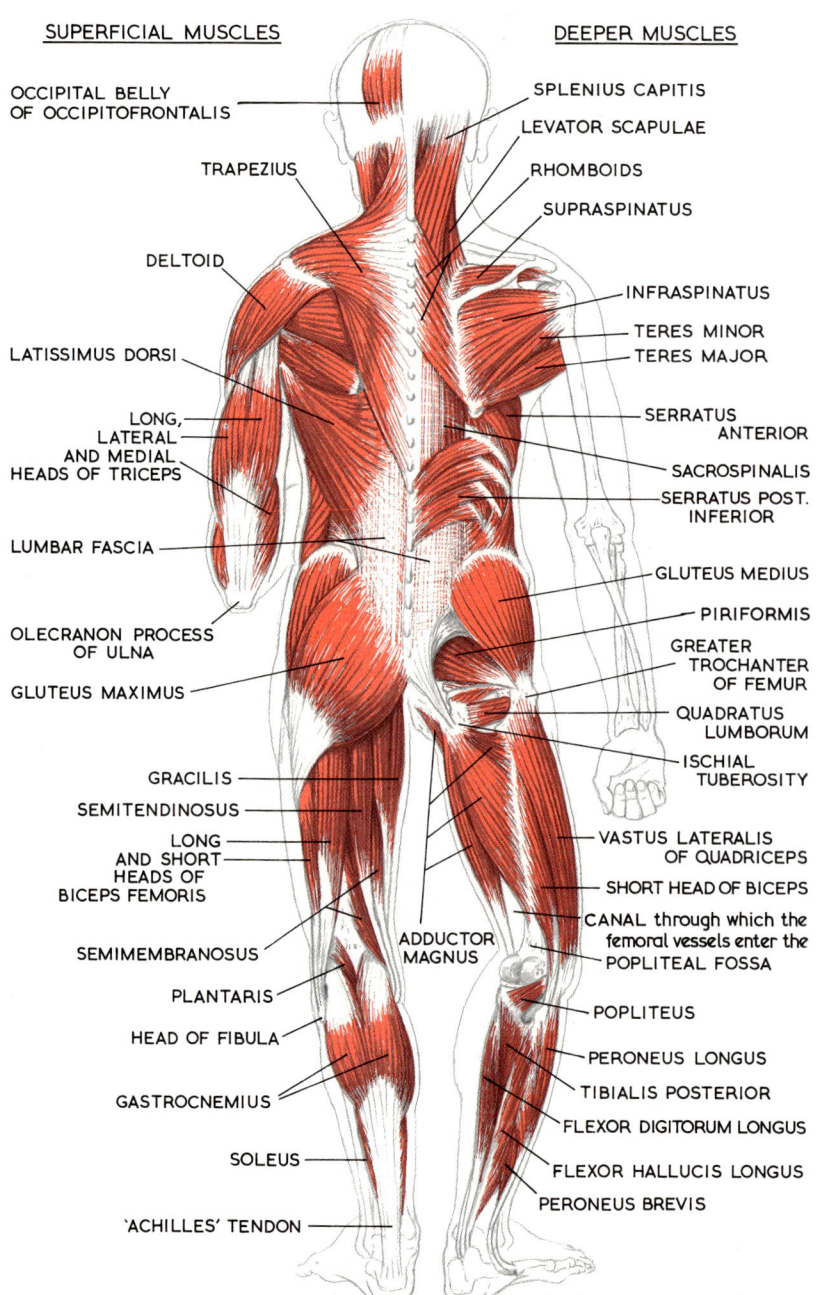

Superficial muscles (left side):
- OCCIPITAL BELLY OF OCCIPITOFRONTALIS
- TRAPEZIUS
- DELTOID
- LATISSIMUS DORSI
- LONG, LATERAL AND MEDIAL HEADS OF TRICEPS
- LUMBAR FASCIA
- OLECRANON PROCESS OF ULNA
- GLUTEUS MAXIMUS
- GRACILIS
- SEMITENDINOSUS
- LONG AND SHORT HEADS OF BICEPS FEMORIS
- SEMIMEMBRANOSUS
- PLANTARIS
- HEAD OF FIBULA
- GASTROCNEMIUS
- SOLEUS
- 'ACHILLES' TENDON

Deeper muscles (right side):
- SPLENIUS CAPITIS
- LEVATOR SCAPULAE
- RHOMBOIDS
- SUPRASPINATUS
- INFRASPINATUS
- TERES MINOR
- TERES MAJOR
- SERRATUS ANTERIOR
- SACROSPINALIS
- SERRATUS POST. INFERIOR
- GLUTEUS MEDIUS
- PIRIFORMIS
- GREATER TROCHANTER OF FEMUR
- QUADRATUS LUMBORUM
- ISCHIAL TUBEROSITY
- VASTUS LATERALIS OF QUADRICEPS
- SHORT HEAD OF BICEPS
- CANAL through which the femoral vessels enter the POPLITEAL FOSSA
- ADDUCTOR MAGNUS
- POPLITEUS
- PERONEUS LONGUS
- TIBIALIS POSTERIOR
- FLEXOR DIGITORUM LONGUS
- FLEXOR HALLUCIS LONGUS
- PERONEUS BREVIS

8 Muscles from behind (superficial ones on the left, deeper ones on the right). It also shows the bones of the right forearm pronated.

PLATES 9

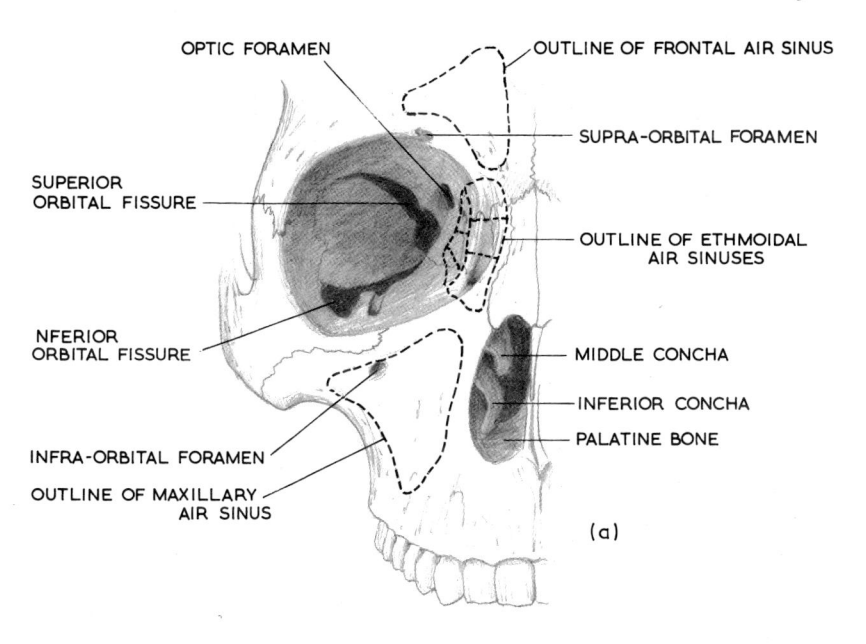

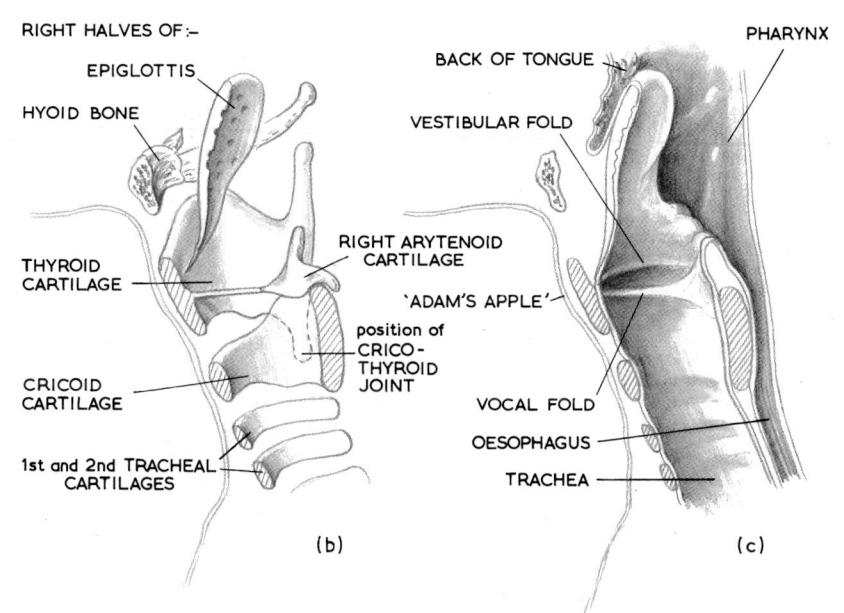

9 (a) The position of the paranasal air sinuses.
(b) The right halves of the laryngeal cartilages.
(c) The right half of the larynx.

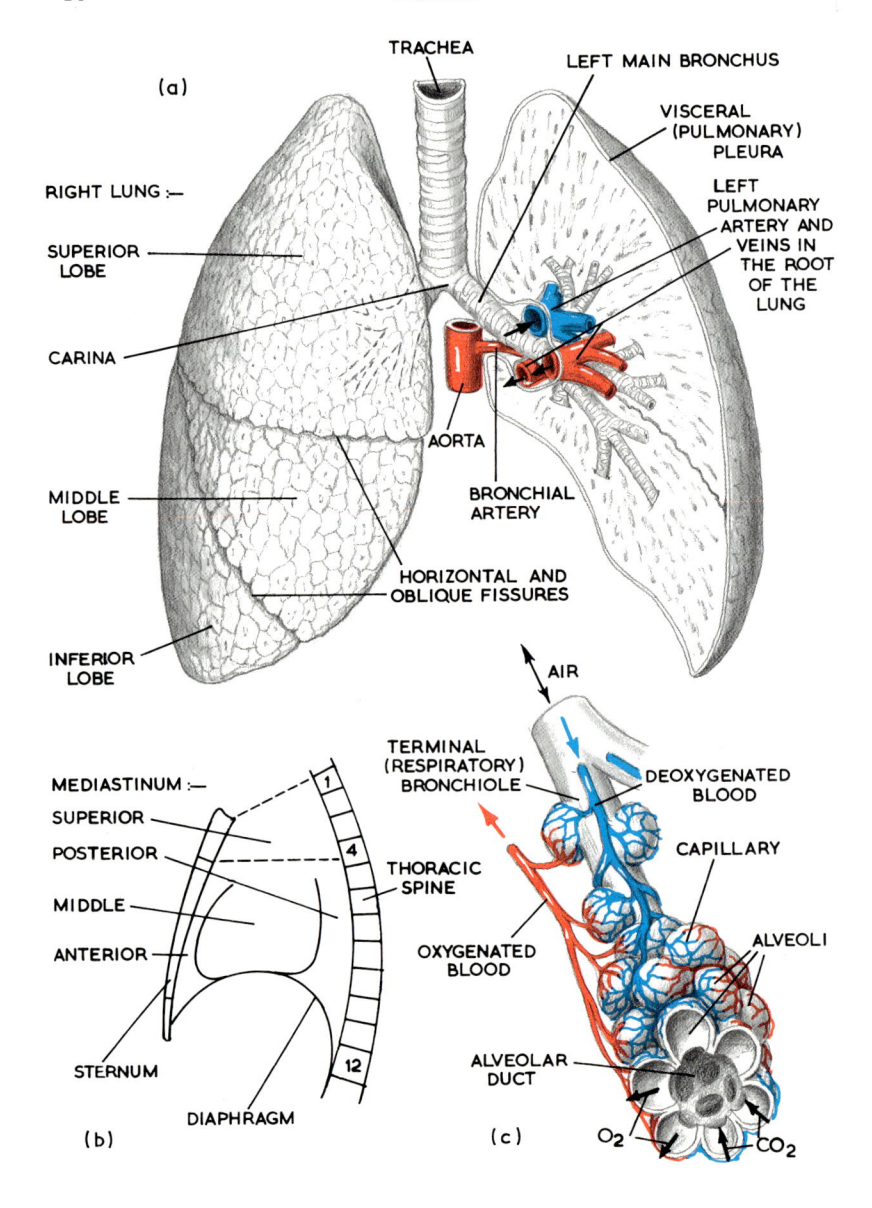

10 (a) The trachea, bronchi and lungs (anterior view, left lung cut open).
(b) The divisions of the mediastinum (diagrammatic).
(c) A pulmonary unit (where gas exchange takes place).

PLATES 11

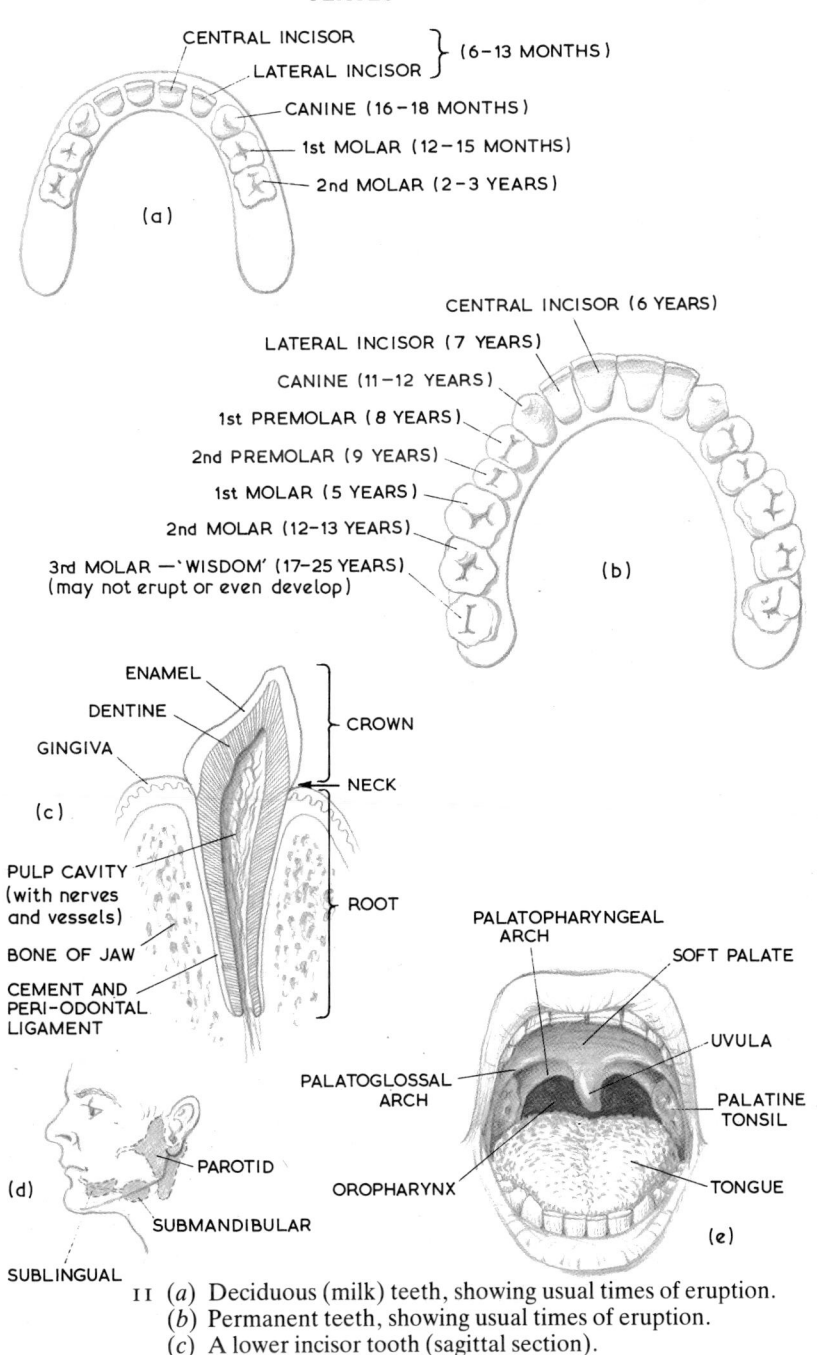

11 (a) Deciduous (milk) teeth, showing usual times of eruption.
(b) Permanent teeth, showing usual times of eruption.
(c) A lower incisor tooth (sagittal section).
(d) The salivary glands.
(e) The mouth cavity from in front.

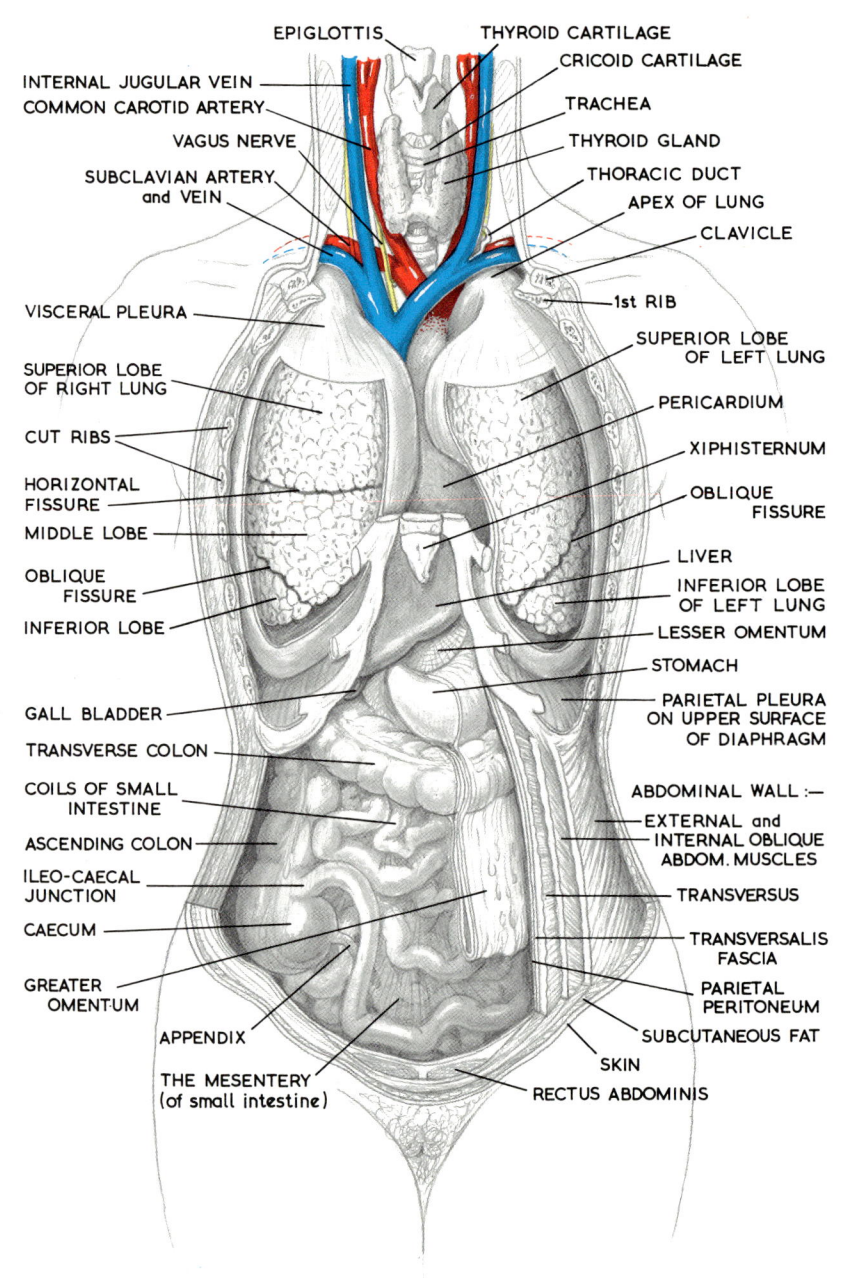

12 The thorax and abdomen from the front: superficial structures.

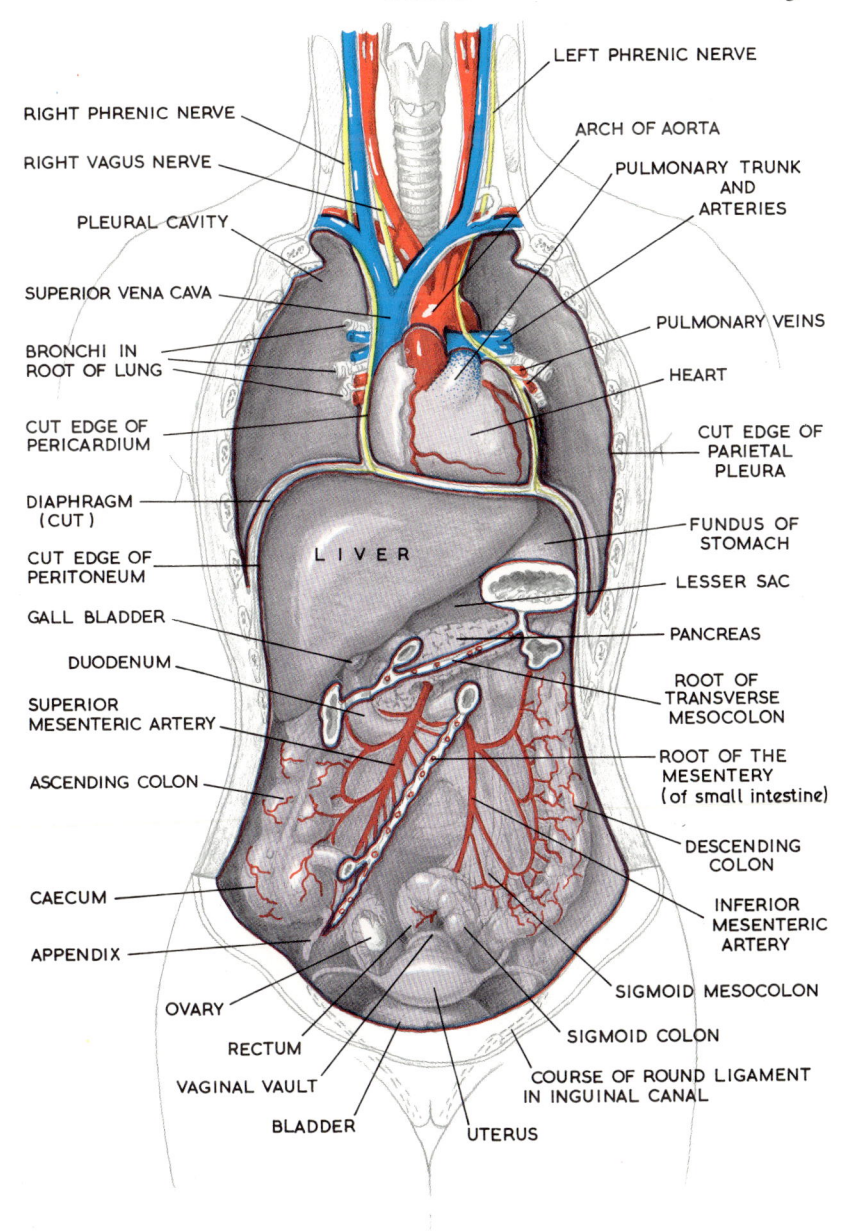

PLATES 13

13 The thorax and abdomen: deeper structures.

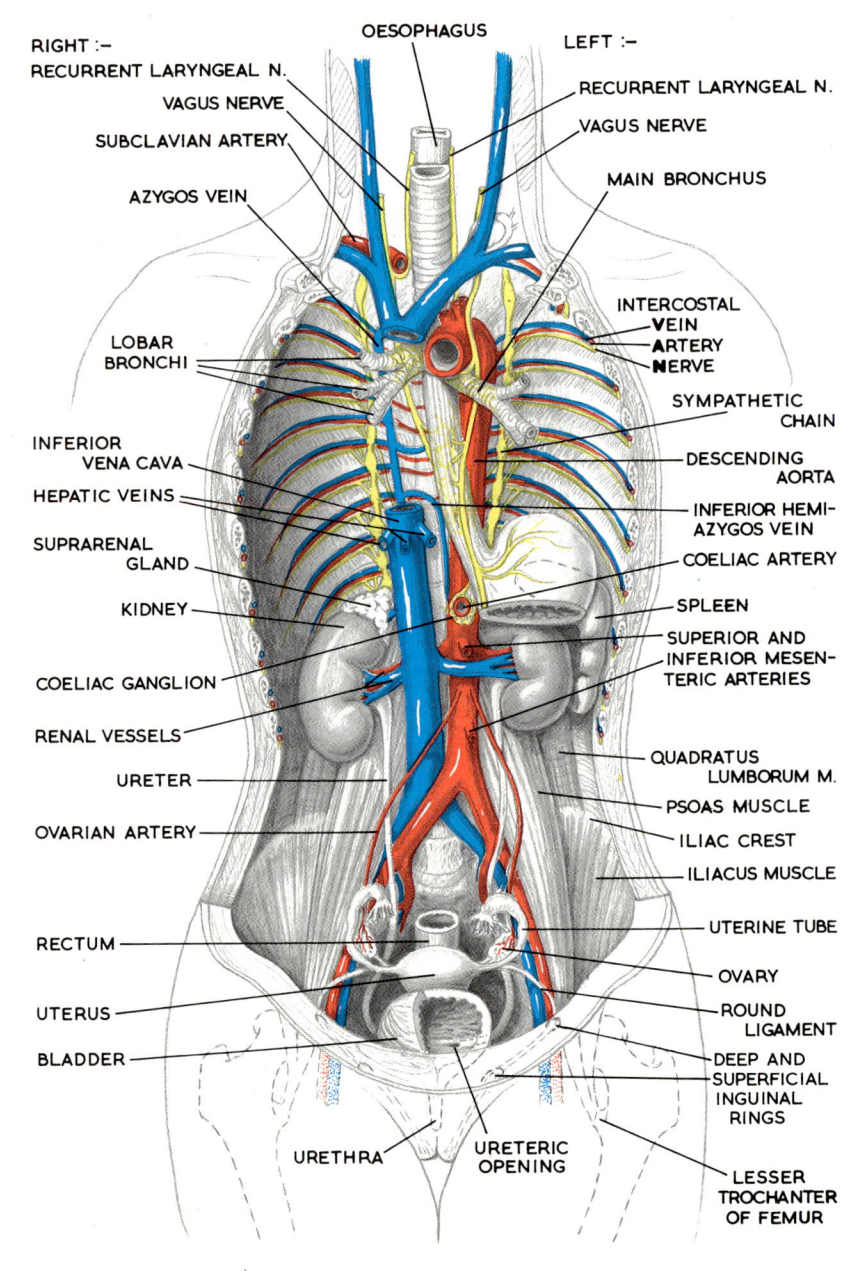

14 The thorax and abdomen: structures on the posterior wall and pelvic contents (female).

PLATES 15

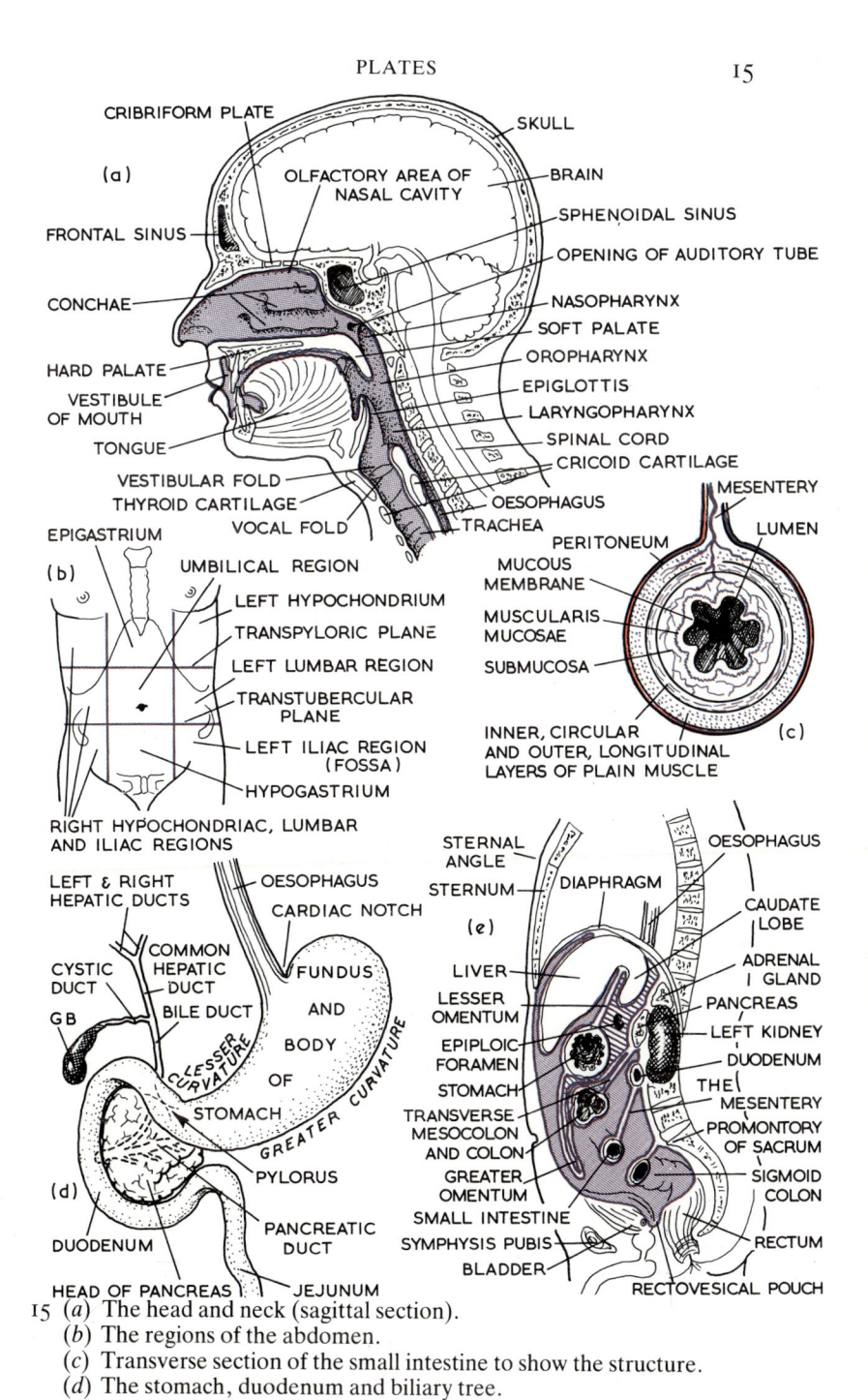

15 (a) The head and neck (sagittal section).
  (b) The regions of the abdomen.
  (c) Transverse section of the small intestine to show the structure.
  (d) The stomach, duodenum and biliary tree.
  (e) Sagittal section of the abdominal cavity to show the peritoneum; the lesser sac is cross-hatched, the greater sac is plain.

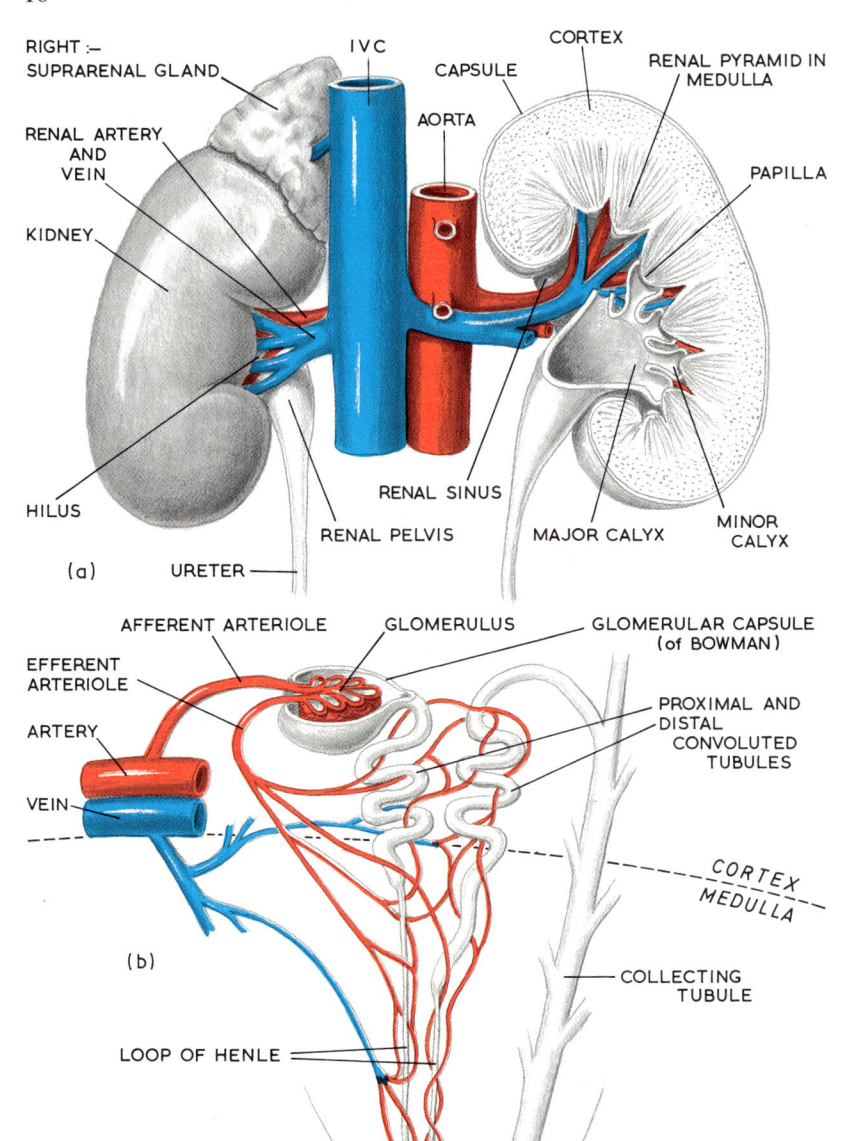

16 (a) The right kidney, with its suprarenal gland, shown complete; the left kidney is sectioned.
(b) A nephron (diagrammatic).

PLATES

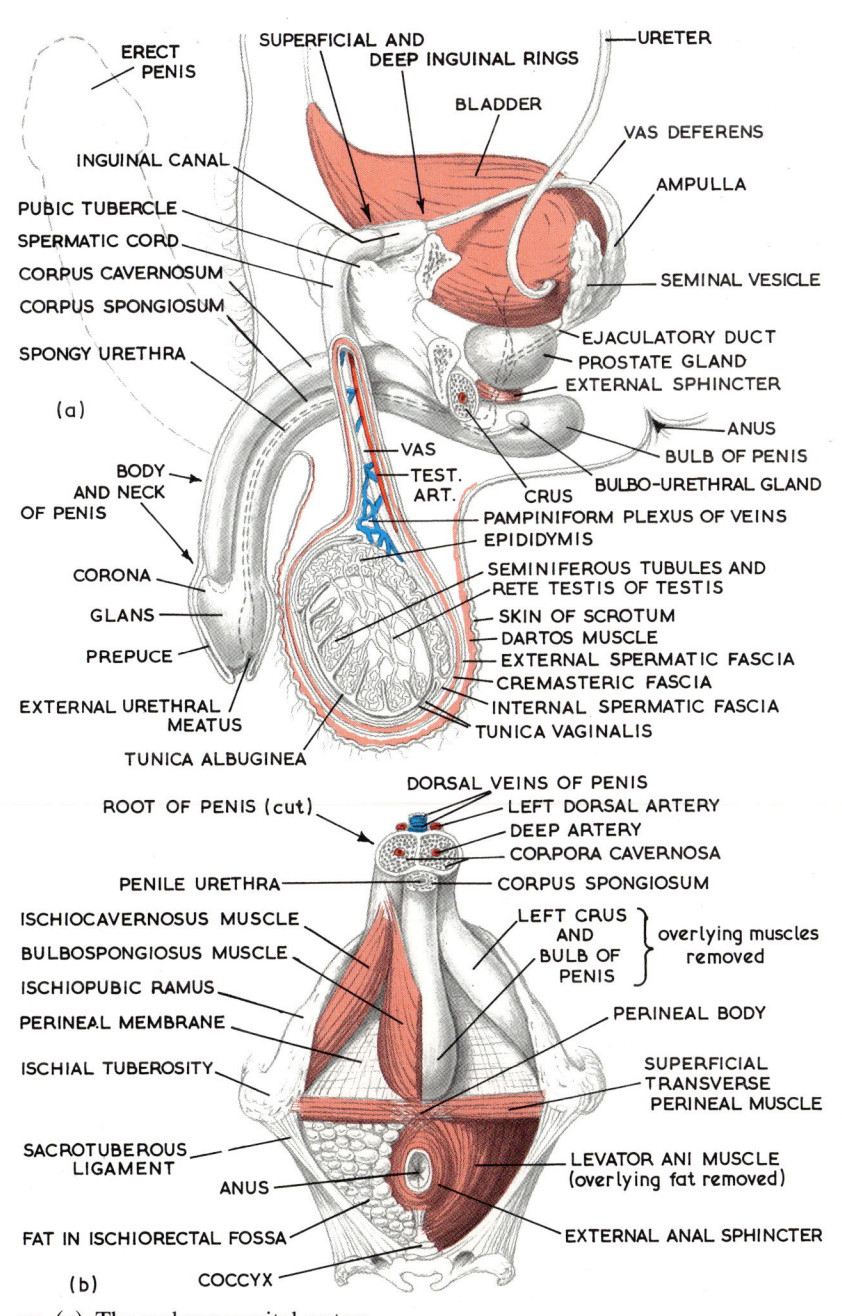

17 (a) The male urogenital system.
(b) The male perineum; the penis is sectioned to show its structure.

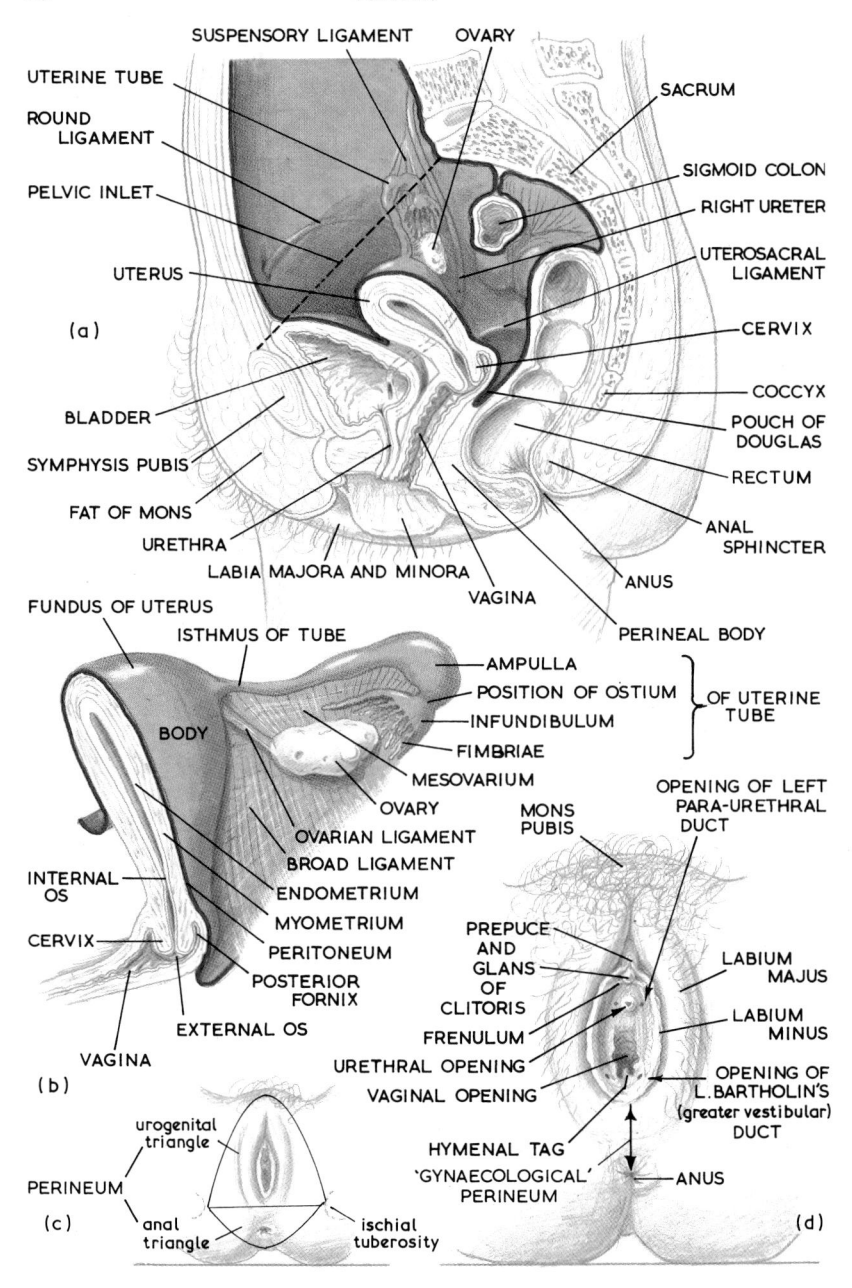

18 The female urogenital system:
   (a) Sagittal section of the pelvis.
   (b) The right half of the uterus and vagina.
   (c) The perineal triangles.
   (d) The vulva.

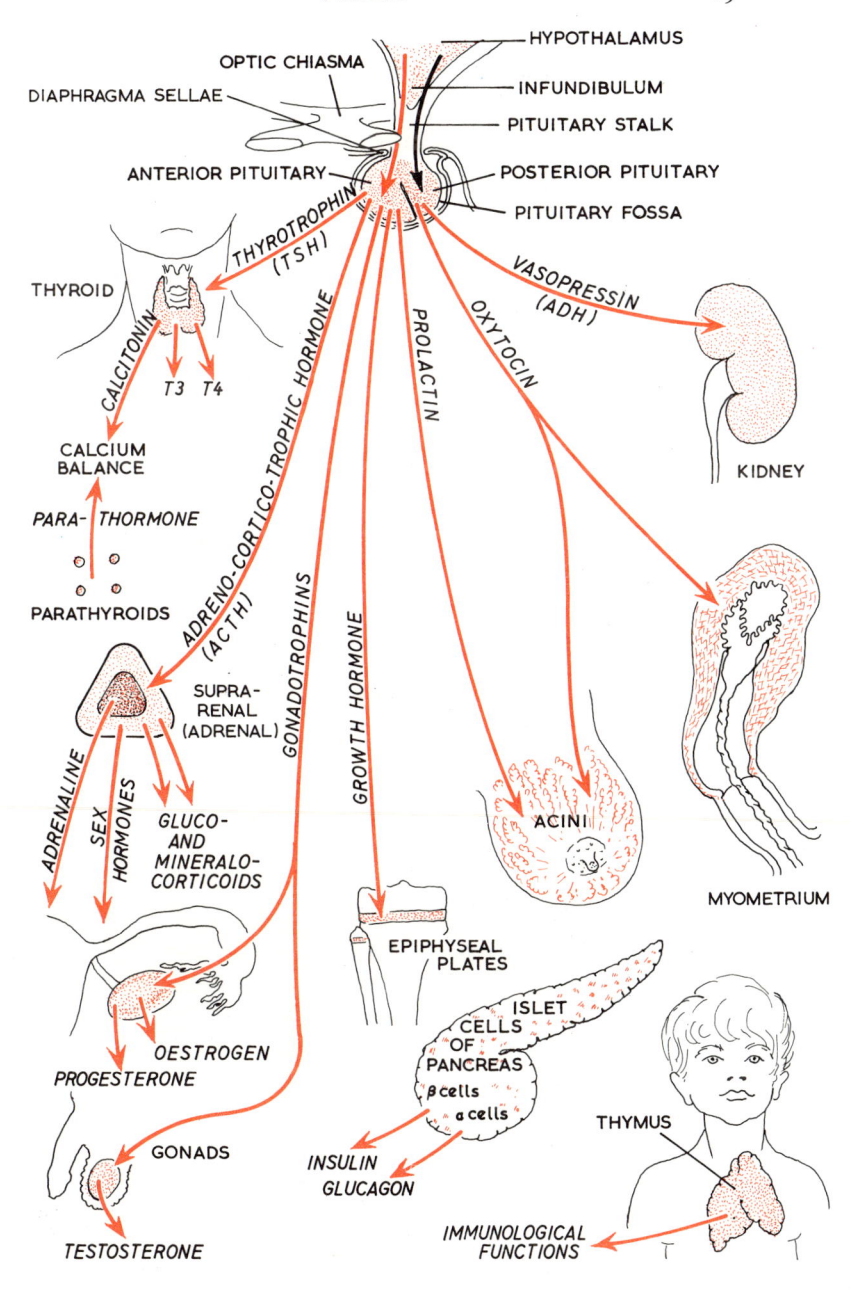

19 Some endocrine organs and their hormones (not to scale).

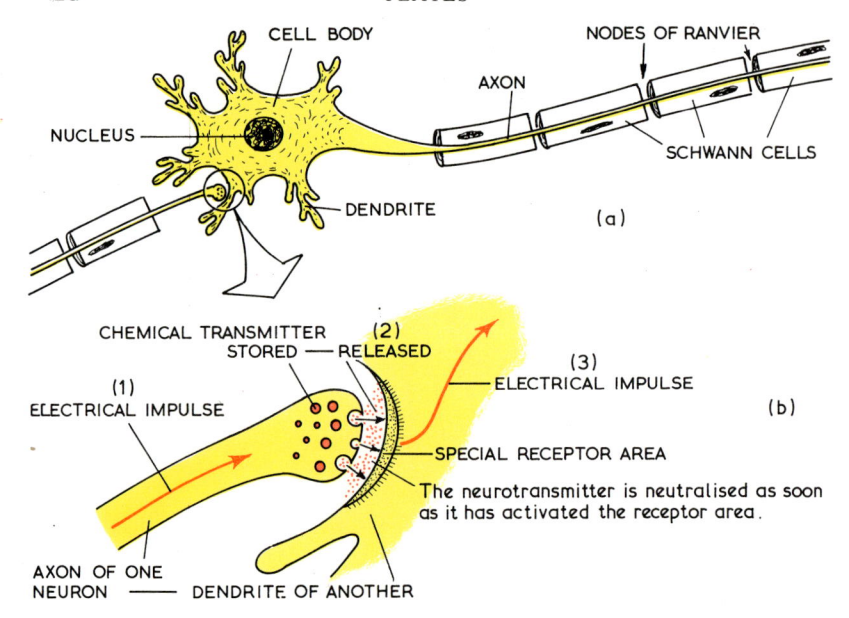

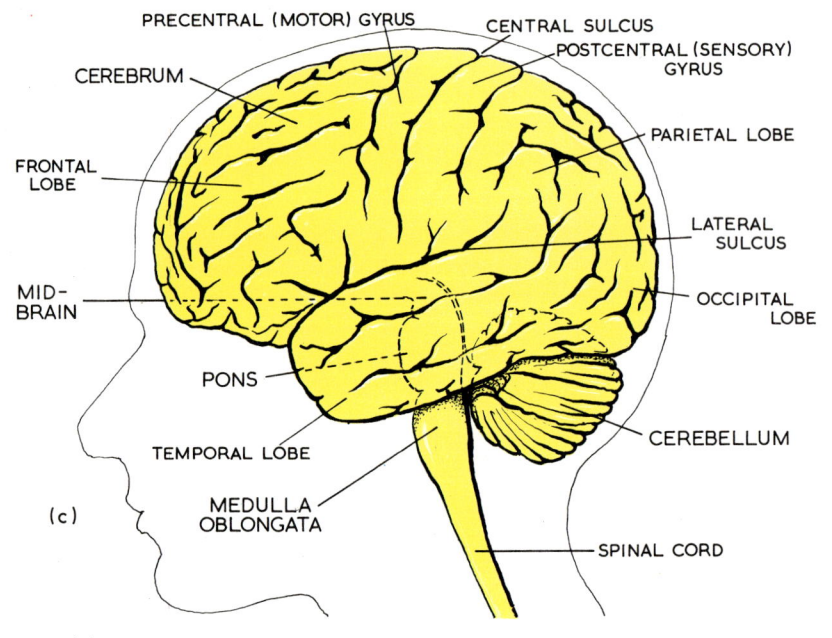

20 (*a*) A neuron.
   (*b*) Detail of a synapse.
   (*c*) The left side of the brain (semidiagrammatic).

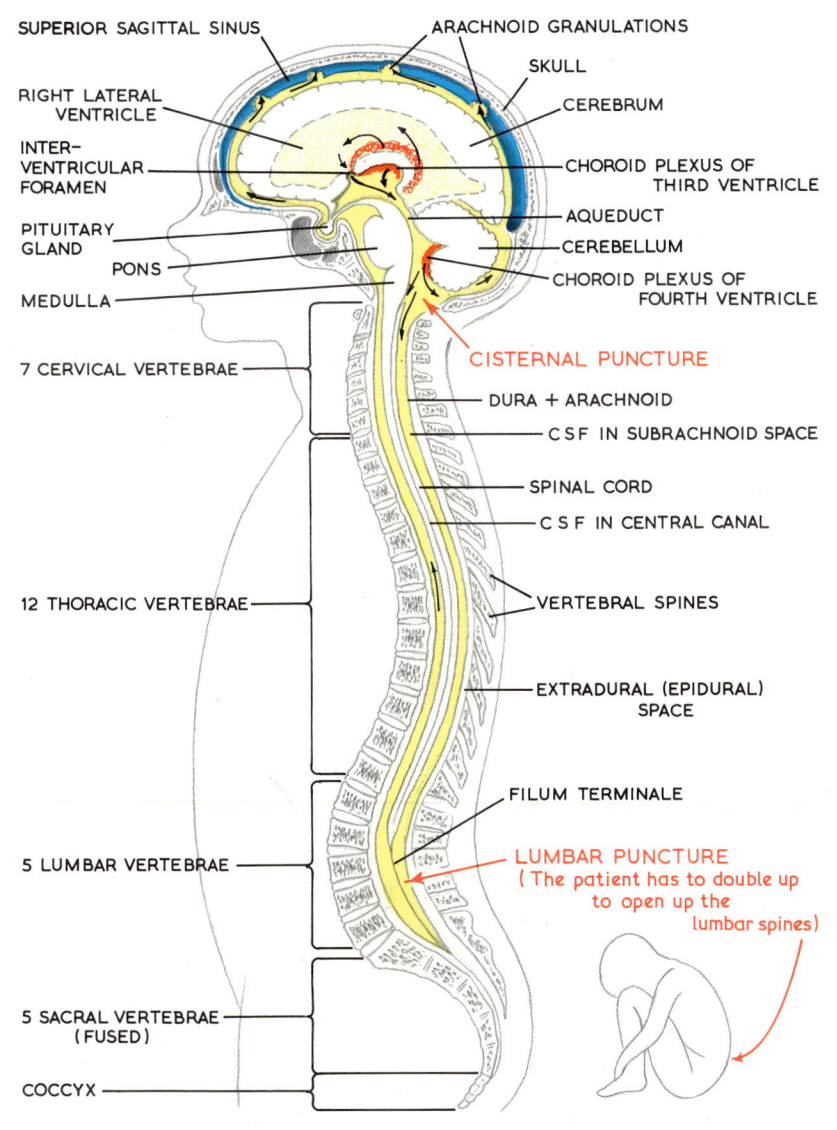

21 Sagittal section of the central nervous system showing the circulation of the CSF (semidiagrammatic).

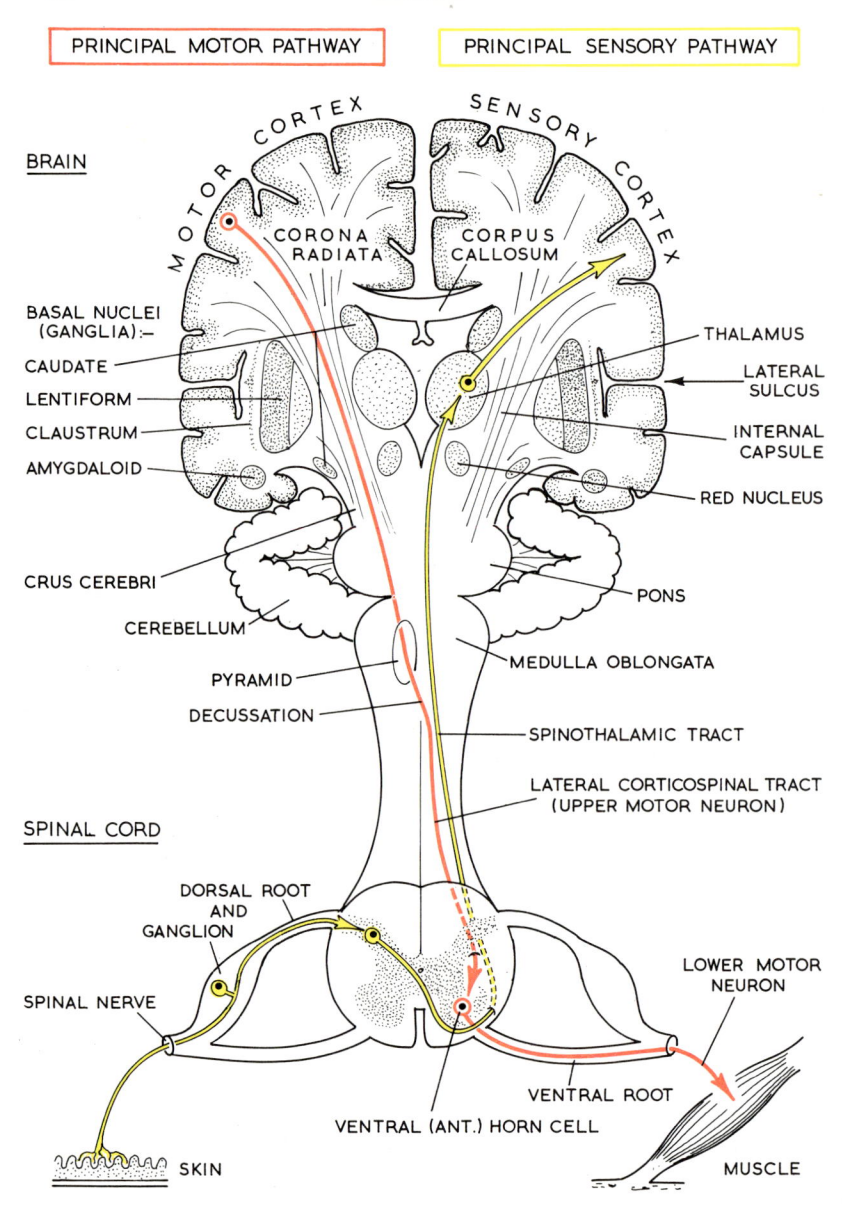

22 The central nervous system: principal motor and sensory pathways (diagrammatic).

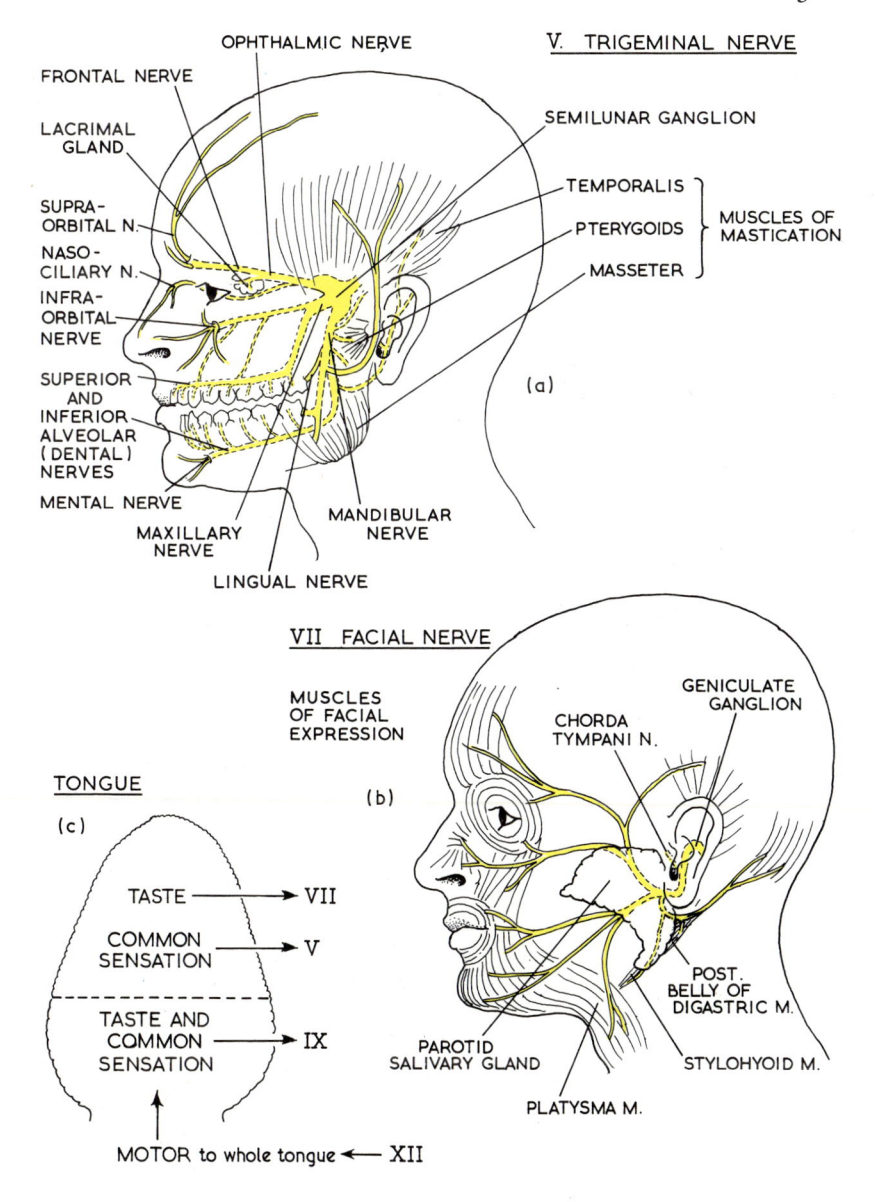

23 The distribution of some of the branches of (*a*) the trigeminal nerve and (*b*) the facial nerve. (*c*) The nerve supply of the tongue.

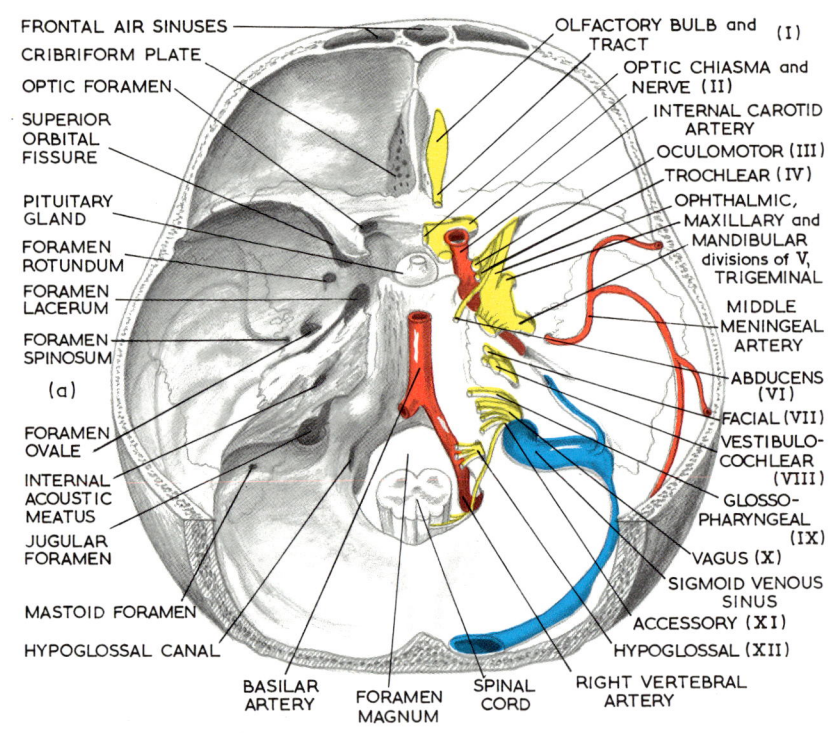

24 (a) The inside of the base of the skull showing foramina (left side) and structures passing through them (right side).

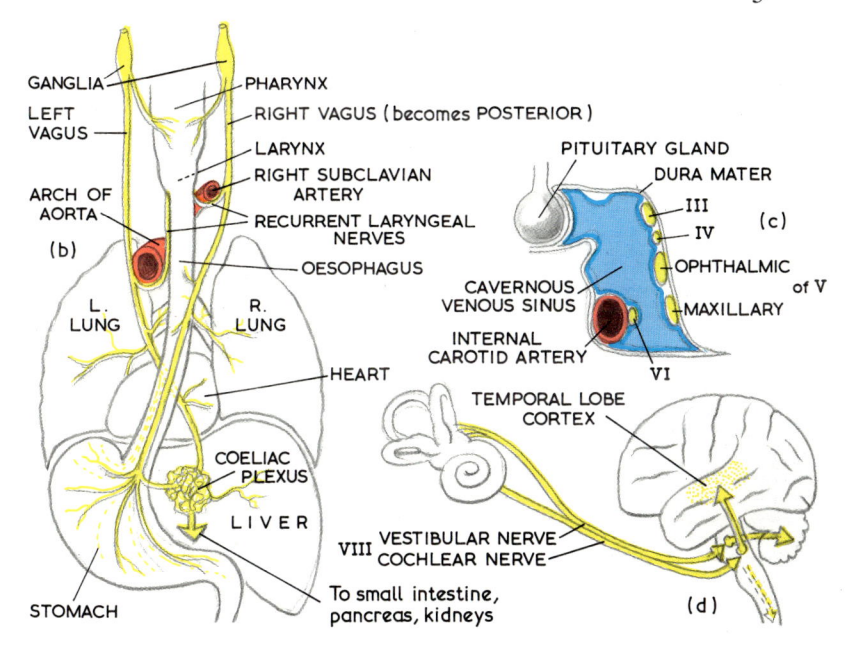

(b) The course and distribution of the vagus nerves from behind (semidiagrammatic).
(c) The cavernous sinus (coronal section).
(d) The central connections of the vestibulocochlear nerve (diagrammatic).

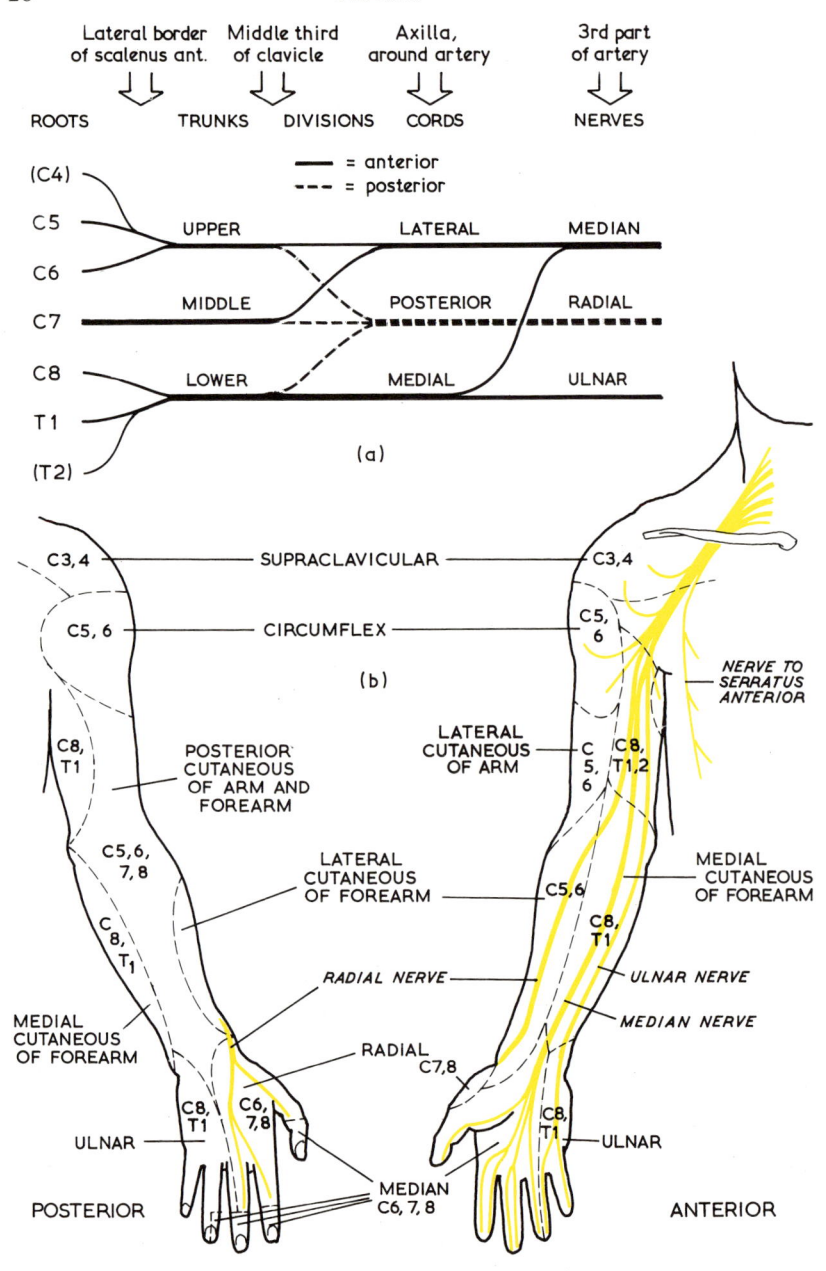

25 (a) Plan of the brachial plexus.
  (b) Cutaneous nerve supply of the right upper limb and course of the main nerves.

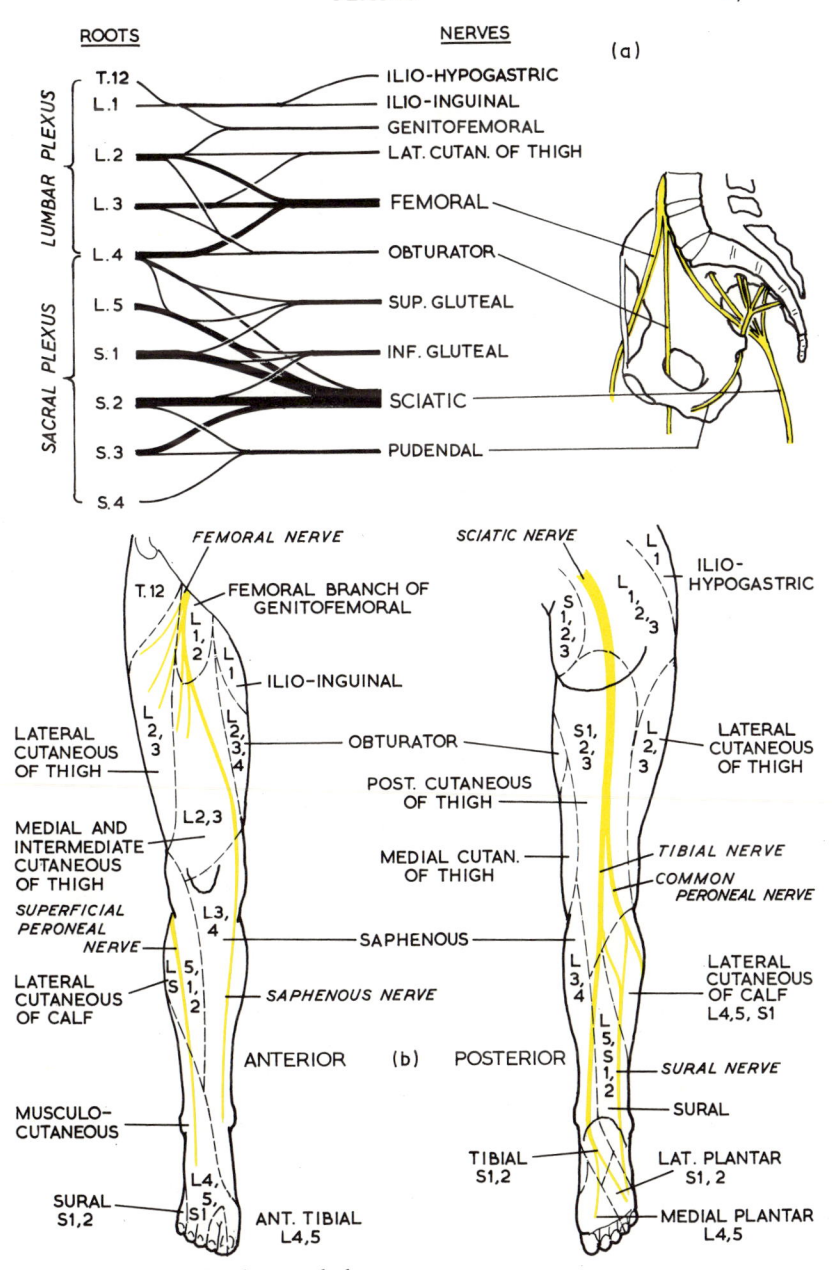

PLATES

26 (a) Plan of the lumbosacral plexus.
 (b) Cutaneous nerve supply of the right lower limb and course of the main nerves.

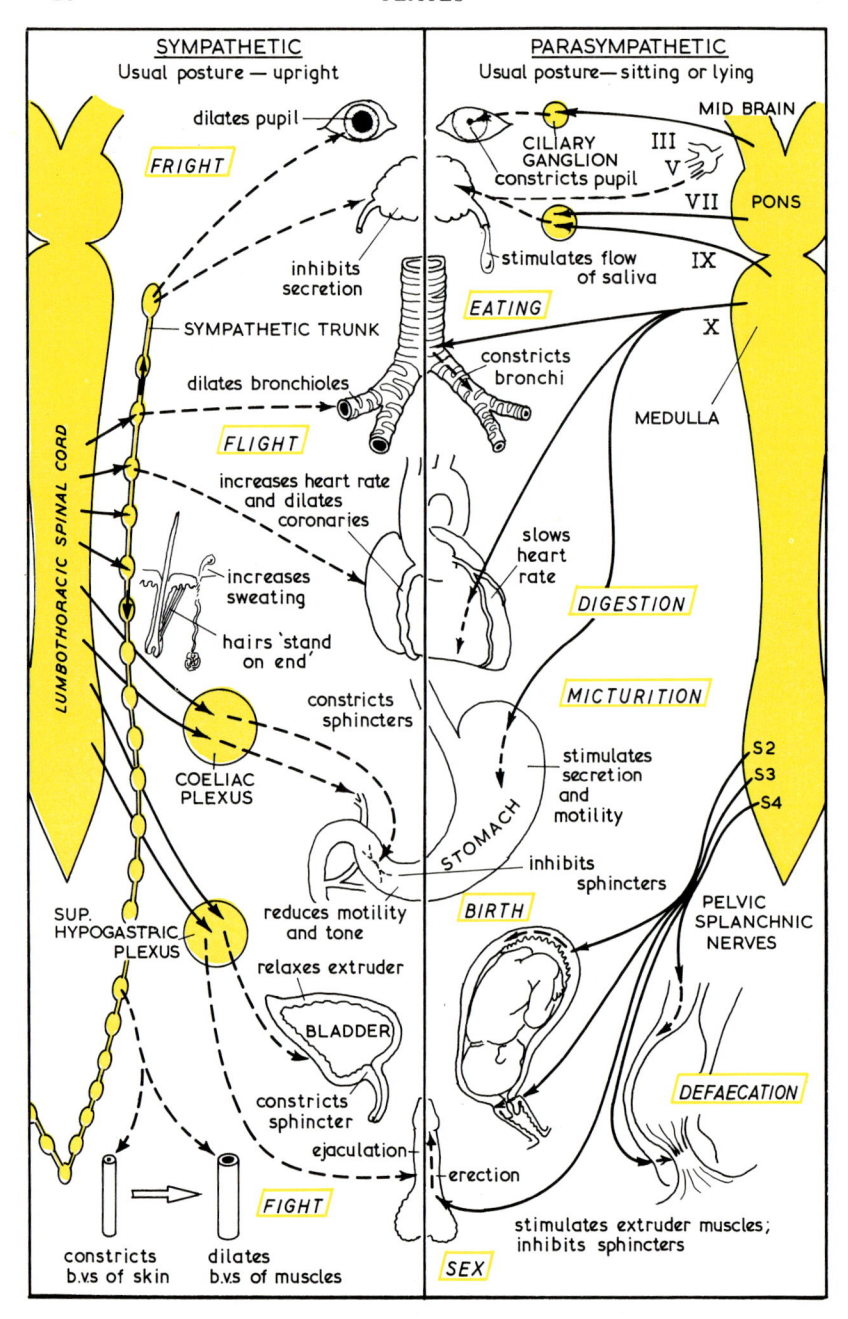

27 The functions of the autonomic nervous system.

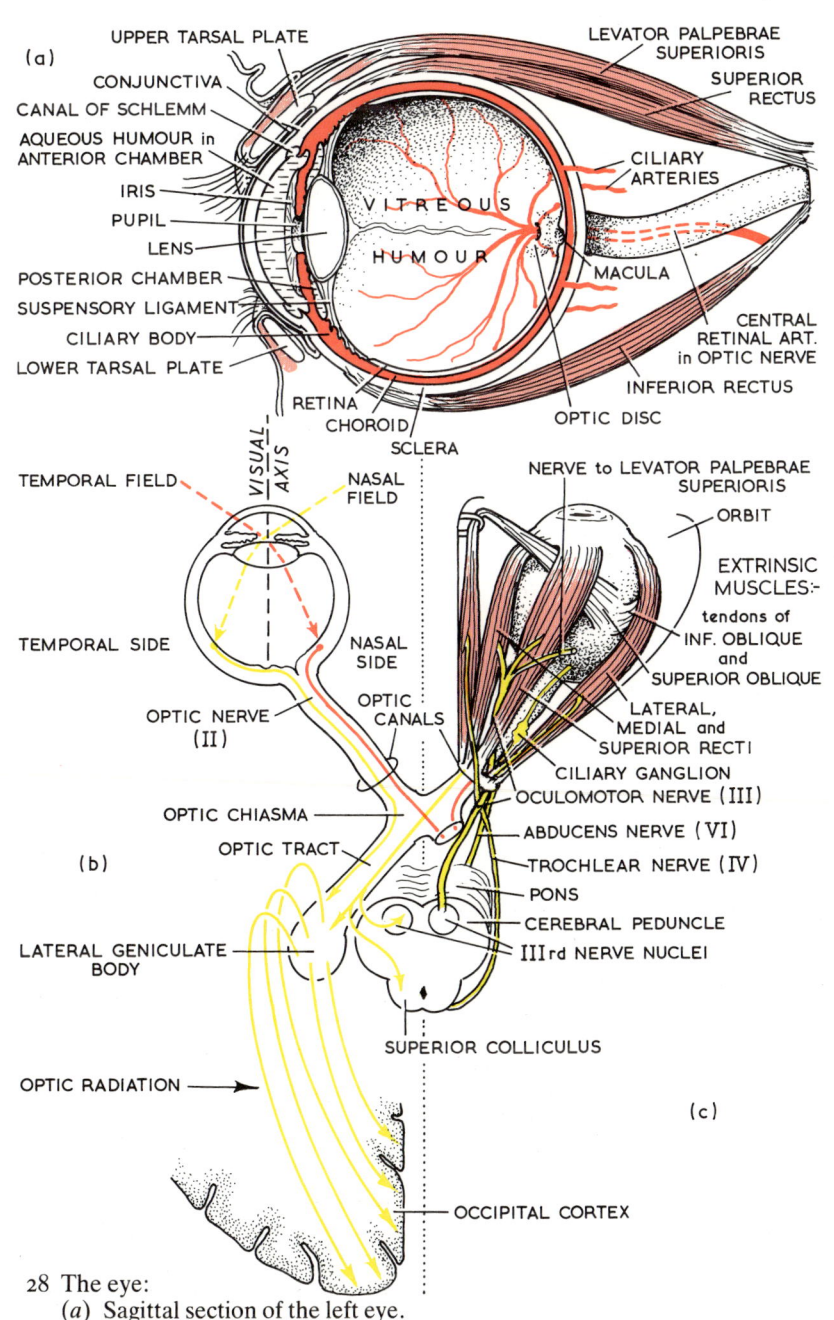

28 The eye:
  (a) Sagittal section of the left eye.
  (b) The optic pathway.
  (c) The extrinsic eye muscles and their nerves.

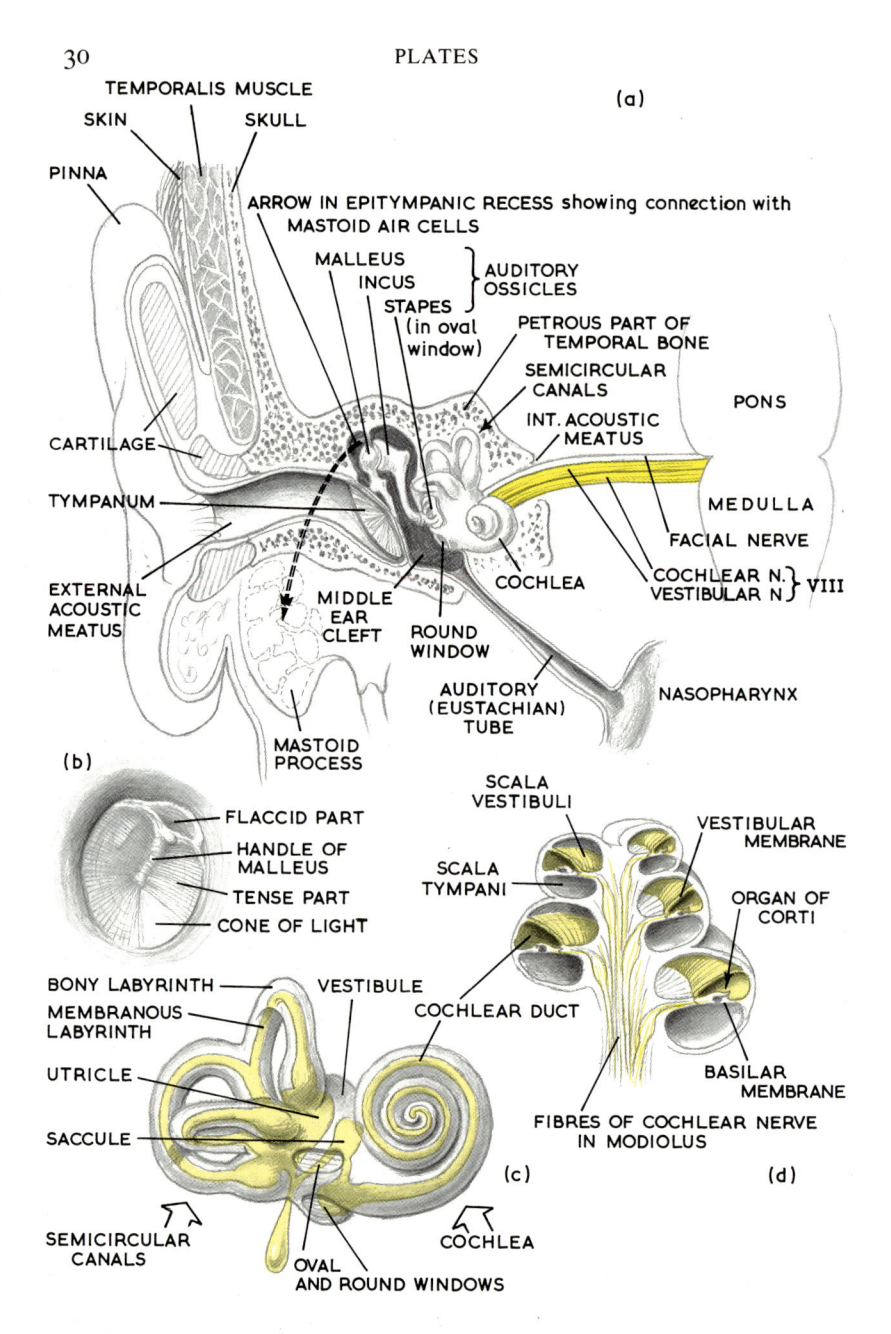

29 The ear:
  (*a*) Coronal section showing external, middle and inner ear (from the front).
  (*b*) Right tympanic membrane, as seen through an auriscope.
  (*c*) The right inner ear.
  (*d*) Section through the cochlea.

PLATES 31

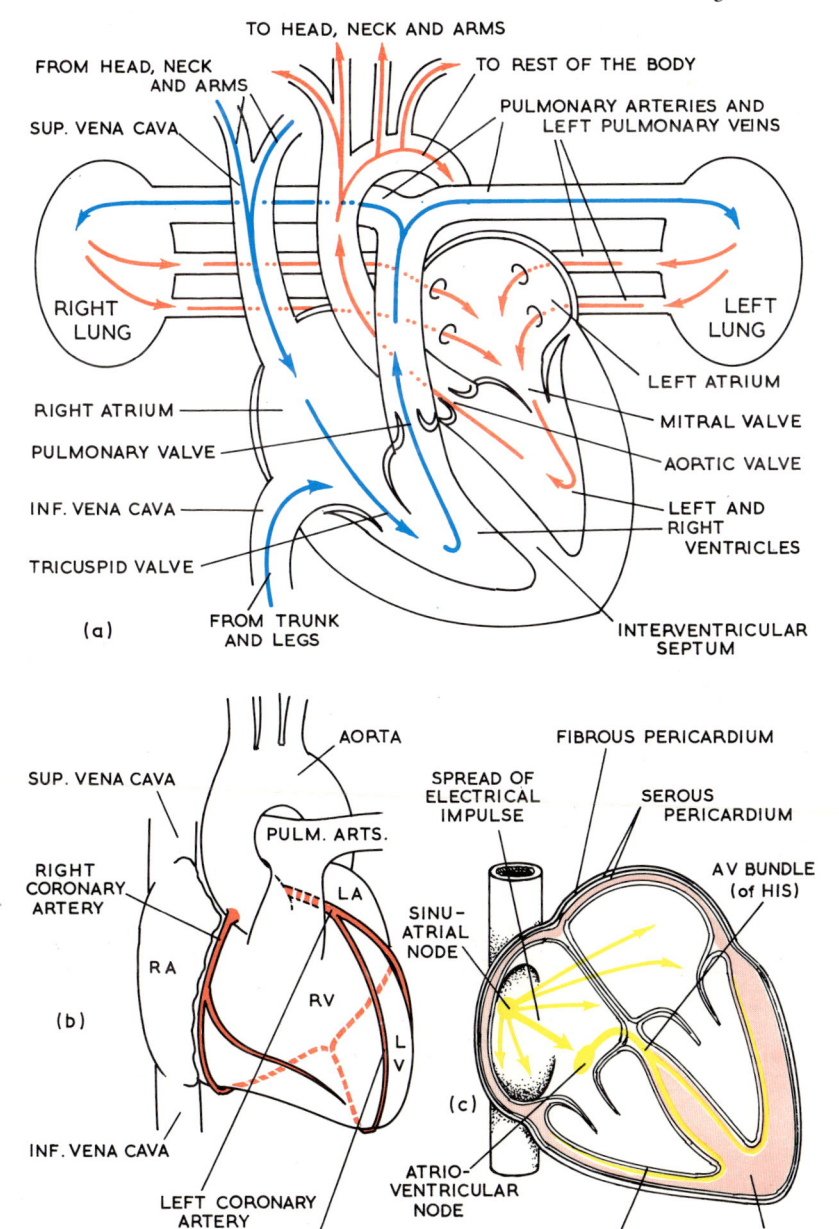

30 (a) The circulation of the blood through the heart and great vessels; red indicates oxygenated blood, blue denotes deoxygenated blood.
(b) The coronary arteries.
(c) Electrical conduction in the heart.

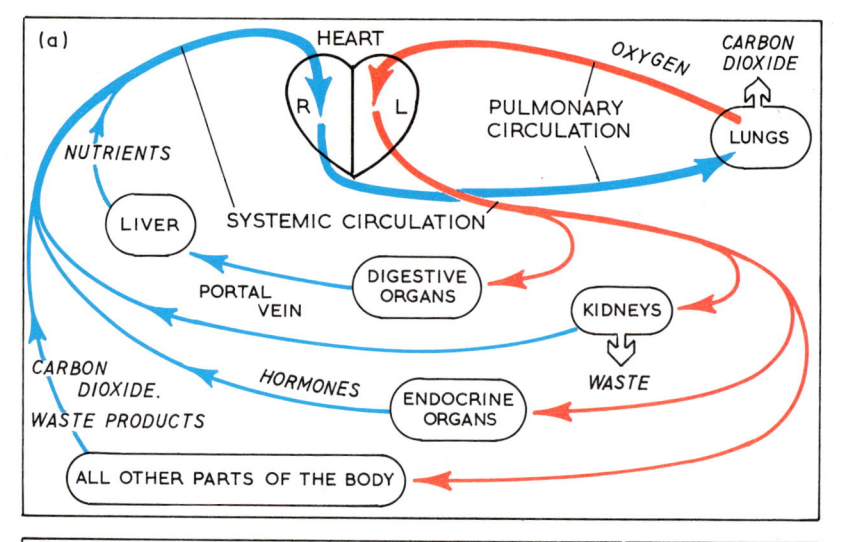

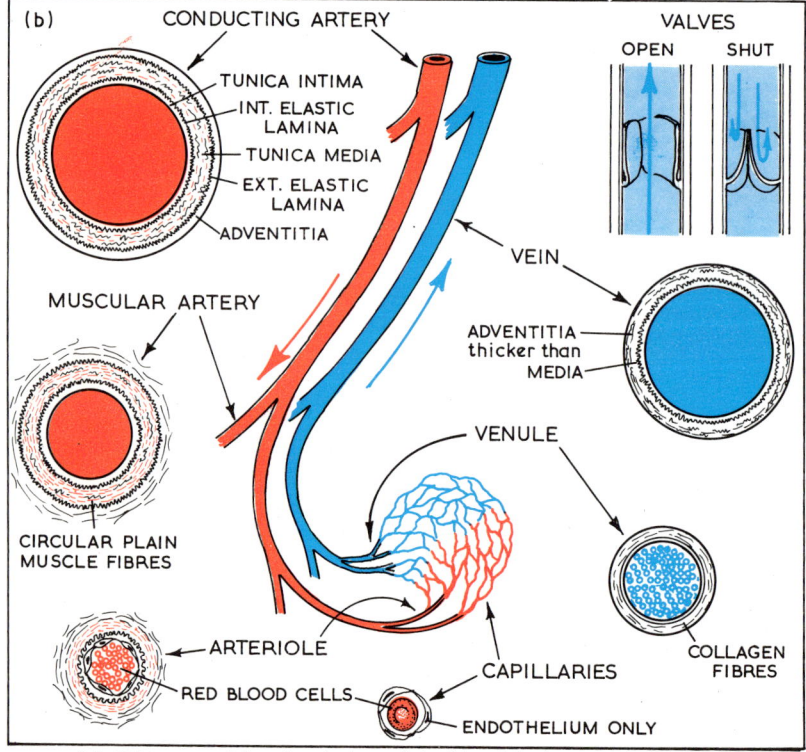

31 (a) Diagram of the circulation of the blood (colours as in Fig. 30).
(b) Blood vessels.

PLATES 33

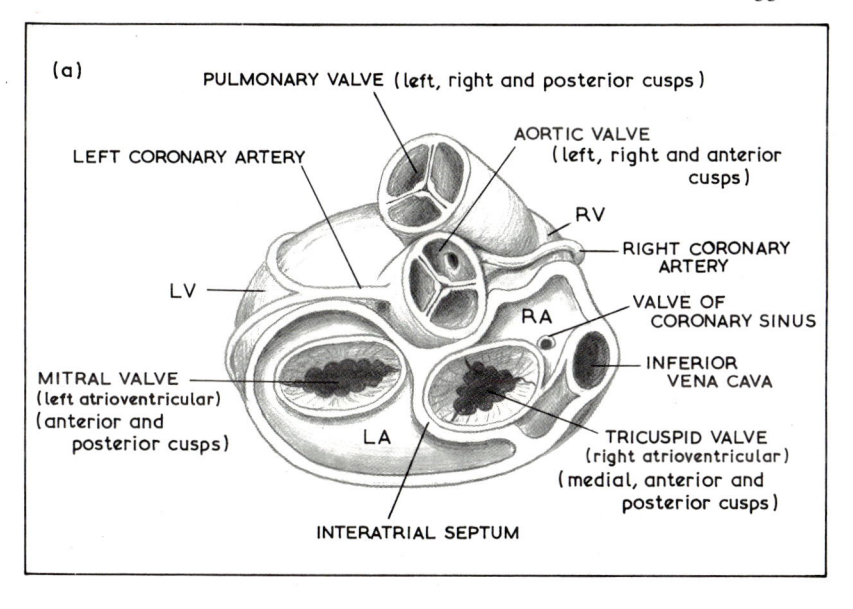

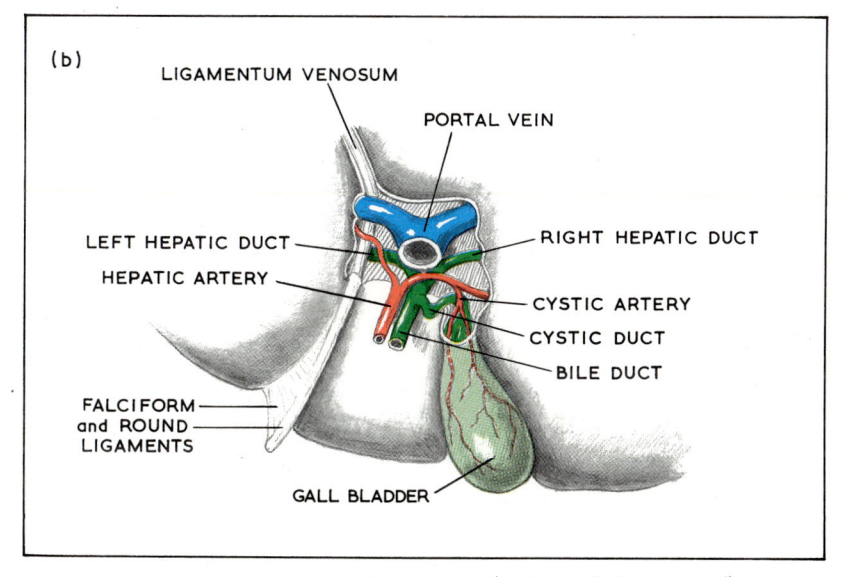

32 (*a*) The heart valves, viewed from above (atria partially removed).
   LA, RA: left atrium, right atrium;
   LV, RV: left ventricle, right ventricle.
(*b*) The porta hepatis.

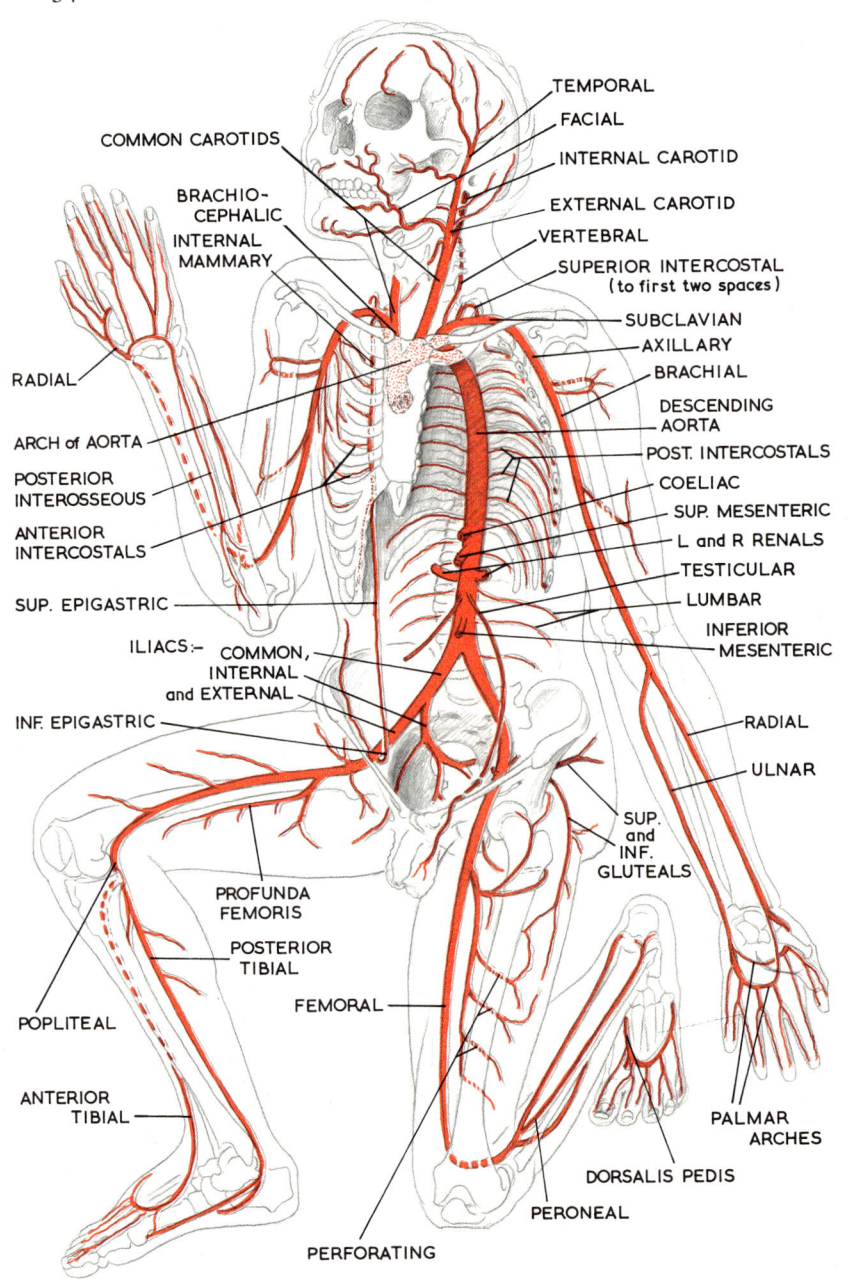

33 Principal arteries.

PLATES

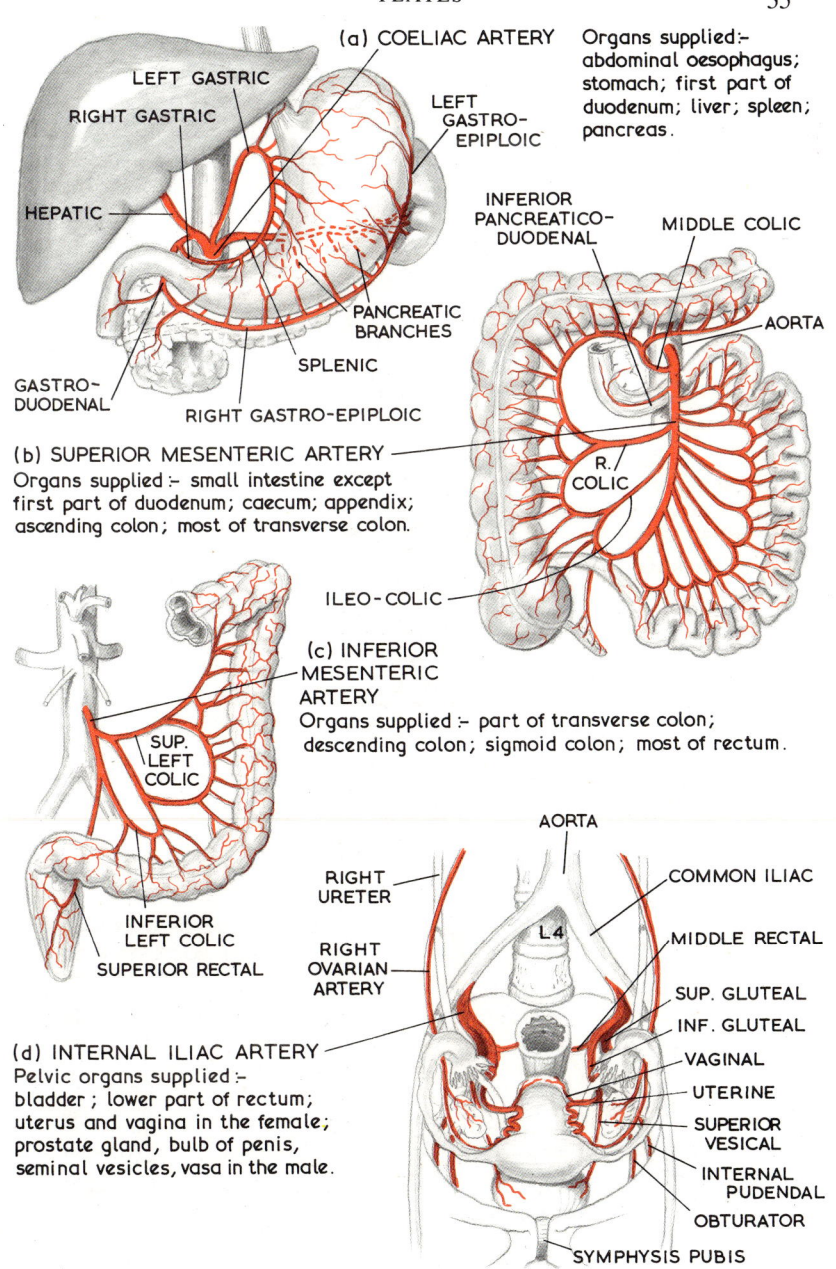

(a) COELIAC ARTERY — Organs supplied:- abdominal oesophagus; stomach; first part of duodenum; liver; spleen; pancreas.

Labels: LEFT GASTRIC, RIGHT GASTRIC, HEPATIC, LEFT GASTRO-EPIPLOIC, INFERIOR PANCREATICO-DUODENAL, PANCREATIC BRANCHES, SPLENIC, GASTRO-DUODENAL, RIGHT GASTRO-EPIPLOIC, MIDDLE COLIC, AORTA, R. COLIC

(b) SUPERIOR MESENTERIC ARTERY — Organs supplied :- small intestine except first part of duodenum; caecum; appendix; ascending colon; most of transverse colon.

Labels: ILEO-COLIC

(c) INFERIOR MESENTERIC ARTERY — Organs supplied :- part of transverse colon; descending colon; sigmoid colon; most of rectum.

Labels: SUP. LEFT COLIC, INFERIOR LEFT COLIC, SUPERIOR RECTAL

(d) INTERNAL ILIAC ARTERY — Pelvic organs supplied :- bladder ; lower part of rectum; uterus and vagina in the female; prostate gland, bulb of penis, seminal vesicles, vasa in the male.

Labels: AORTA, RIGHT URETER, RIGHT OVARIAN ARTERY, L4, COMMON ILIAC, MIDDLE RECTAL, SUP. GLUTEAL, INF. GLUTEAL, VAGINAL, UTERINE, SUPERIOR VESICAL, INTERNAL PUDENDAL, OBTURATOR, SYMPHYSIS PUBIS

34 The arterial supply to the abdominal and pelvic viscera.
    (*a*) Coeliac artery.
    (*b*) Superior mesenteric artery.
    (*c*) Inferior mesenteric artery.
    (*d*) Internal iliac artery.

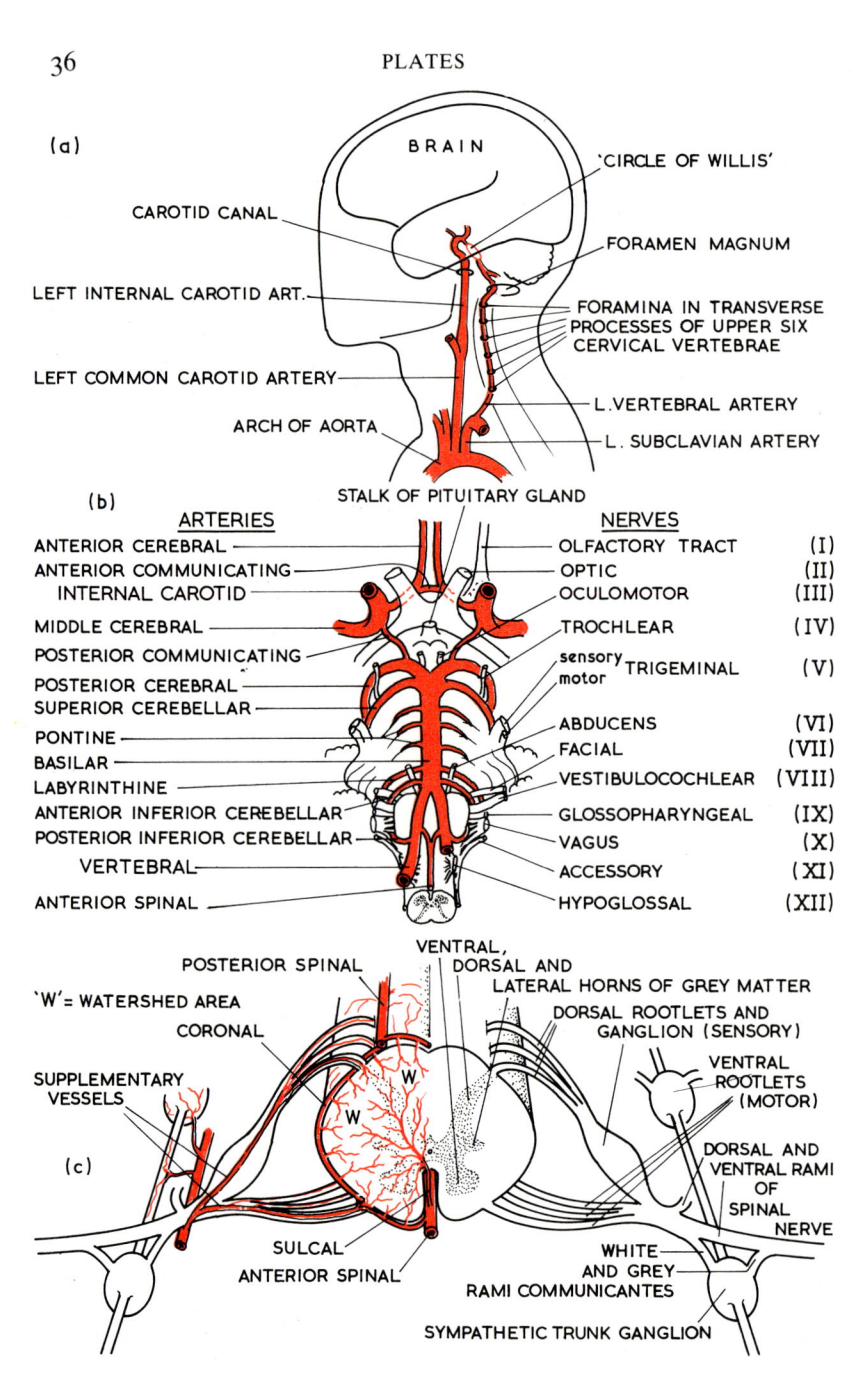

35 The blood supply to the central nervous system:
    (*a*) Sources of arterial blood to the brain (diagrammatic).
    (*b*) The brain stem from below showing arteries and cranial nerves.
    (*c*) The spinal cord and sympathetic chains (blood supply shown on the left).

PLATES 37

36 Principal veins.

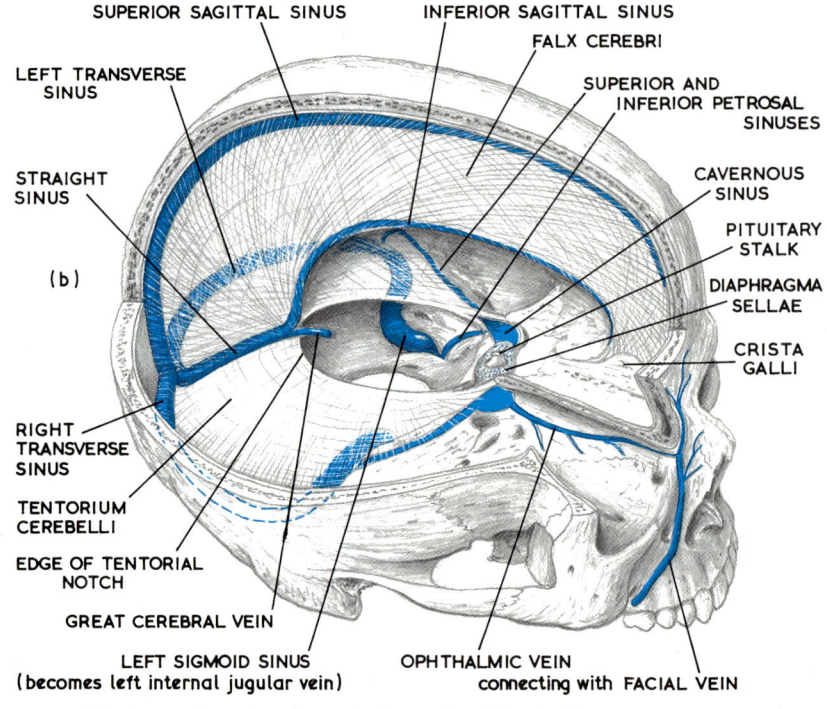

37 (*a*) Coronal section through the vault of the skull.
   (*b*) The dura mater and venous sinuses (part of the skull and the whole of the brain removed).

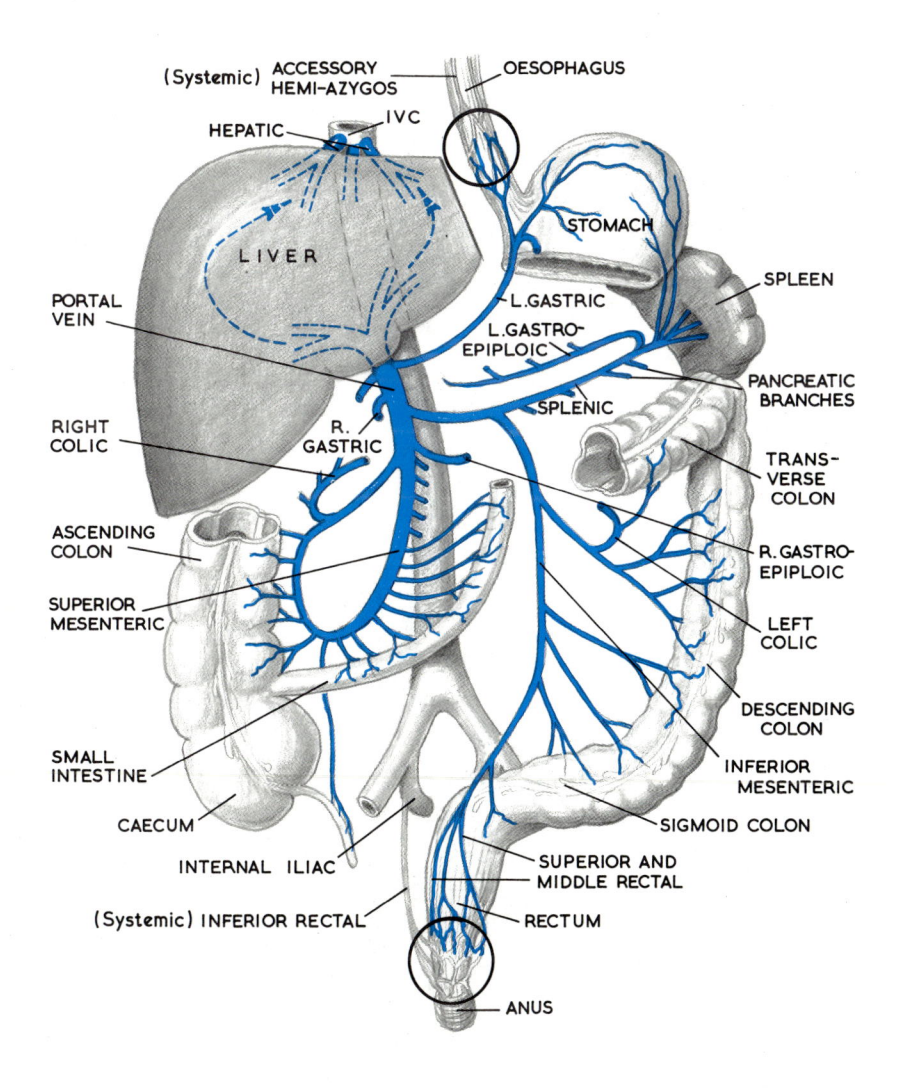

38 The hepatic portal system; important portosystemic anastomoses are ringed.

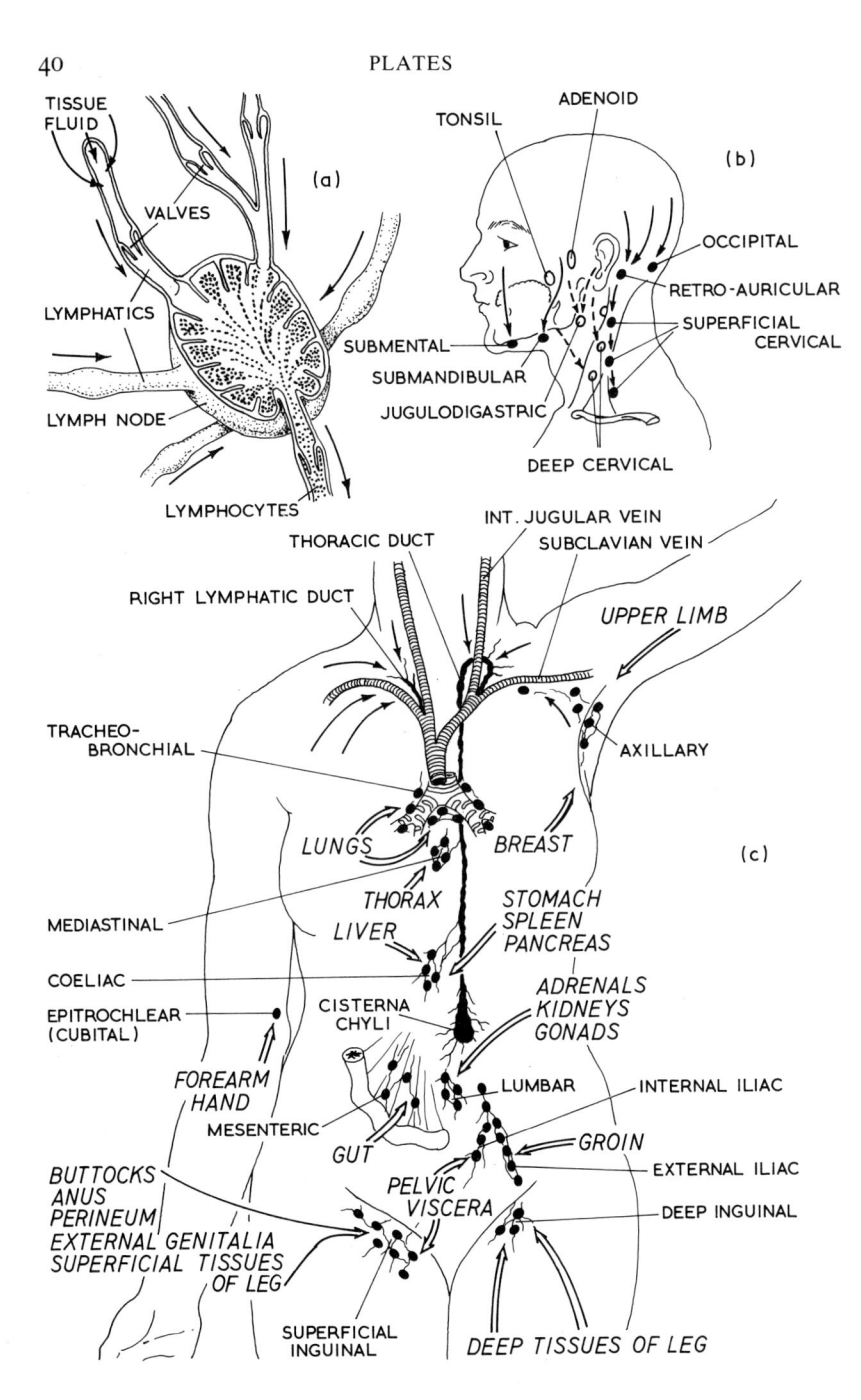

39 (a) Detail of a single lymph node.
(b) Principal nodes and lymphatic drainage of the head and neck.
(c) Principal nodes and drainage of the trunk and limbs.

PLATES

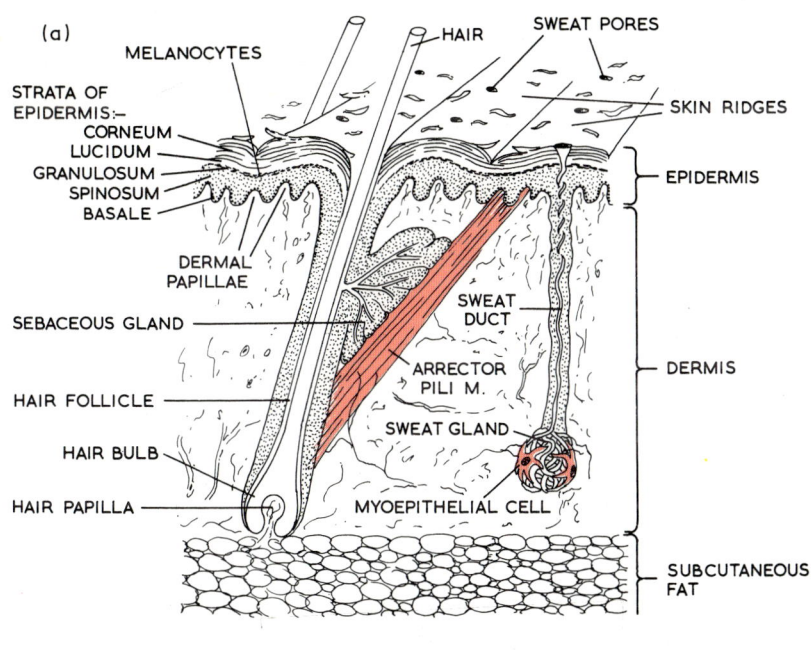

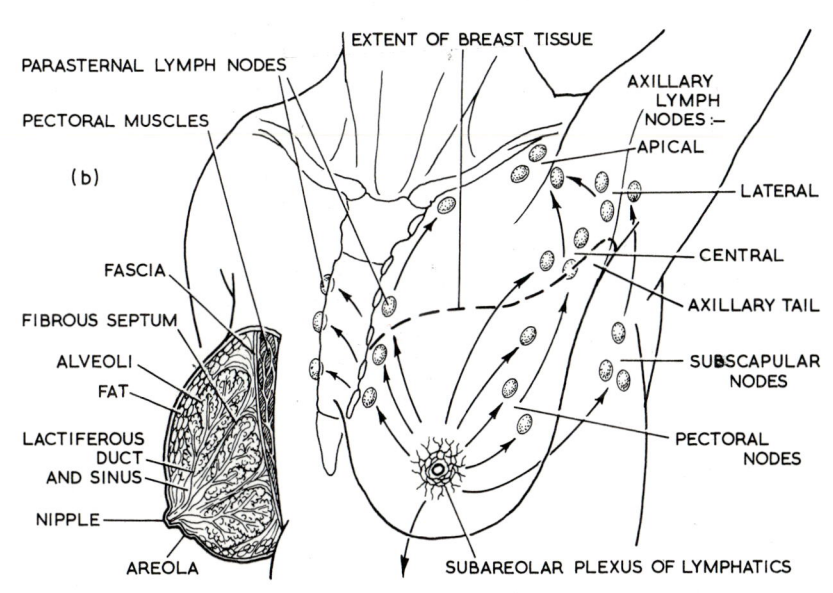

40 (*a*) Section through the skin (much enlarged).
  (*b*) The lactating breast: structure shown in the right breast, lymphatic drainage from the left.

The Skeleton
(Figs 1–6)

BONES

Bones consist of a protein matrix (osteoid) stiffened by calcium and phosphate. They form rigid levers for movement and protect soft structures such as the brain, spinal cord and viscera.

Bones may be (*Fig. 2*):

(1) *long*, e.g., femur, humerus
(2) *short*, e.g., carpals and tarsals
(3) *flat*, e.g., sternum and scapula
(4) *irregular*, e.g., pelvic bones
(5) *sesamoid*, e.g., the patella

A long bone consists of a shaft and two ends. Its outer cortex of dense bone encloses the medullary cavity containing looser cancellous bone, blood vessels and fatty bone marrow. The shaft is covered by membranous periosteum, which contains blood vessels and bone-forming cells.

The Skull (22 bones) (*Figs 2 and 3*)

The bones of the head are tightly connected by fibrous sutures. Apart from the mandible (lower jaw) they are immobile.

The 8 bones of the cranium protect the brain. They are:

1 *frontal* bone
2 *parietal* bones

1 *occipital* bone
2 *temporal* bones
1 *sphenoid* bone, which with the occipital forms most of the base of the skull
1 *ethmoid* bone, between the brain and nasal cavity

The Face (14 bones) (*Fig. 2*)

These form the bony eye sockets (orbits) and the skeleton of the nose, mouth and jaws:

2 *maxillae* (upper jaw and most of the hard palate)
1 *mandible* (lower jaw)
2 *zygomatic* (cheekbones)
2 *nasal* bones (bridge of the nose)
1 *vomer* (bony part of the nasal septum)
2 *lacrimal* bones (medial to the orbits)
2 L-shaped *palatine* bones (side walls of nose, part of roof of mouth)
2 *inferior nasal conchae* or *turbinates*, projecting into the cavity of the nose from its lateral walls

The U-shaped *hyoid* bone lies in the neck but is attached to the styloid process of the temporal bone by the stylohyoid ligament.

The Spine (33 vertebrae) (*Fig. 3*)

The vertebral column protects the spinal cord and provides attachments for muscles. Between the 24 true vertebrae are intervertebral discs of fibrocartilage which allow the spine to flex and extend (see *Fig. 6c*). There are:

7 *cervical* vertebrae. The first is called the *atlas*, the second the *axis*
12 *thoracic* or *dorsal* vertebrae
5 *lumbar* vertebrae
9 *false* vertebrae, which have no intervertebral discs but are fused, so there is no movement between them. They form

the *sacrum* (5) and the *coccyx* (4).

The Thoracic Cage (*Fig. 3*)

This consists of the thoracic vertebrae, the sternum and 12 pairs of ribs.

All ribs articulate posteriorly with the thoracic vertebrae, but their anterior connections vary. The first 7 pairs from above downwards (*true ribs*) are joined to the sternum by their own costal cartilages. The remaining 5 pairs (*false ribs*) are joined by costal cartilages to the cartilage above.

The 11th and 12th (*floating*) ribs have no anterior attachments.

The *sternum* (breastbone), lying in the midline anteriorly, consists of:

(1) the *manubrium*, above
(2) the *body*
(3) the *xiphoid process* at the tip, below

The Shoulder Girdle (4 bones) (*Fig. 3*)

These attach the upper limb to the trunk.

On each side:

> 2 *clavicles* (collar bones) hold the upper limbs away from the torso to allow free movement
> 2 *scapulae* (shoulder blades)

The Upper Limbs (*Fig. 2*)

On each side these have:

> 1 *humerus* (upper arm)
> 1 *radius* ⎫
> 1 *ulna* ⎬ (forearm)
> 8 *carpal* (wrist) bones (*Fig. 4a*)
> 5 *metacarpals* (palm of hand)
> 14 *phalanges* (fingers, 3 each; thumb 2)

The Pelvic Girdle (4 bones) (*Fig. 2*)

The pelvis protects and supports the viscera, provides attachments for the muscles of the abdomen and lower limbs and transmits the weight of the body to the legs.

It is designed for stability as well as movement. On each side there is:

>   1 *innominate* (hip) bone which has 3 parts that meet at the acetabulum (hip joint socket):
>   (*a*) the *ilium* articulates with the sacrum at the sacroiliac joint
>   (*b*) the *ischium*, below and behind
>   (*c*) the *pubis* which meets the pubis of the opposite side anteriorly in the midline at the symphysis pubis

The Lower Limbs (*Fig. 2*)

On each side these have:

>   1 *femur* (thigh bone)
>   1 *tibia* (shin bone)
>   1 *fibula*
>   1 *patella* (kneecap)
>   7 *tarsal* bones (Fig. 4b) (part of ankle and arch of foot)
>   5 *metatarsal* bones of the arch and sole of the foot
>   14 *phalanges*, forming the toes

JOINTS (*Fig. 6*)

Joints are formed where bones articulate with each other.

Joints allowing free movement are *synovial* joints (*Fig. 5b*). *Bursae* are pockets of synovial membrane placed around some joints. Intra-articular cartilage is found in, for example, the knee joint (*menisci* or *semilunar cartilages*).

Synovial Joints

These are of 6 kinds (*Fig. 6b*):
(1) *Ball-and-socket* joints, for example the hip and shoulder, allowing movement in all directions.
(2) *Hinge* joints, for example the elbow joint, interphalangeal joints, where the articular surfaces are moulded to allow flexion and extension only.
(3) *Condylar* joints, for example the knee and the temporomandibular joint. These are like hinge joints but some additional movement is possible because the joint surfaces have condyles (are knobbly) rather than conforming closely to each other as in true hinge joints.
(4) *Ellipsoid* joints: an oval, convex surface fits into a concave socket to allow flexion, extension, adduction, abduction and circumduction, for example the radiocarpal joint at the wrist and the joints between the metacarpals and the phalanges.
(5) *Saddle* joints: the surfaces are concavoconvex, like a saddle, to allow free movement, for example the first carpometacarpal joint at the base of the thumb.
(6) *Plane* joints, whose articular surfaces are nearly flat, and whose limited gliding movement is restricted by ligaments, for example the intervertebral joints and those between the individual carpal and tarsal bones.

Joints allowing little or no movement

Fibrous joints, for example the skull sutures, the peg-and-socket joints at the insertion of the teeth roots into the jaw, and where an interosseous ligament unites the tibia and fibula.

Cartilaginous joints, for example the symphysis pubis, the intervertebral discs, the sternomanubrial joint.

Muscles
(*Figs 7–8*)

Muscles can contract to produce movement. There are three types:

Striped

Striped (striated) muscle is used for willed (voluntary) movement. The striped muscle fibres lie parallel to each other and are gathered into bundles sheathed in *connective tissue*. The muscle is attached to bone at its origin and insertion by tough cord-like *tendons* or by *aponeuroses* (flat tendons). Muscle contraction is initiated by nerve impulses.

Unstriped

Unstriped (plain or involuntary) muscle is found chiefly in the walls of the hollow viscera such as the gut, uterus and bladder, in the walls of the blood vessels and in the iris. It consists of long, thin spindle-shaped cells. Smooth muscle is not under voluntary control and its contraction is a slow, sustained squeezing.

Cardiac

Cardiac muscle, found only in the heart, is striped though less definitely than voluntary muscle. Cardiac muscle is not under voluntary control. Its contraction is controlled by nerves but it can also initiate its own contractions. Some cardiac muscle fibres (*Purkinje fibres*) conduct electrical impulses through the heart.

The Respiratory System
(*Figs 9 and 10*)

This consists of the nose, pharynx, larynx, trachea, bronchi, bronchioles and lungs. Its functions are ventilation and gas exchange; air is transmitted to and from the alveoli of the lungs, where oxygen is absorbed into the blood for use in metabolism and the waste product, carbon dioxide, is excreted back from the blood into the expired air.

THE NOSE

The *external nose* is a pyramidal structure supported by a framework of bone and cartilage. The *nasal cavity* lies within it.

At its root the nose joins the forehead, and the nasal bones underlie the bridge. The nasal cavity is divided into two parts by the *nasal septum*, bony above and cartilaginous below. Just above the nostril is a dent in the septum called the *vestibule*. This is covered with skin and coarse hairs. Above this, the septum is covered with mucous membrane. The lower, *respiratory* area is thick, spongy and vascular and contains many mucous glands. Above it is the *olfactory* area, which extends onto the roof of the cavity and the superior concha and contains scent receptors.

The posterior apertures of the nasal cavity, *choanae*, open into the *nasopharynx*. The lateral wall has three *conchae (turbinates)*, curled plates of bone covered by thick mucous membrane and projecting into the cavity. Below the conchae lie the openings from the paranasal sinuses and nasolacrimal duct.

The *paranasal sinuses* (*Figs 9a and 15a*) are cavities in the bones of the skull around the nasal cavity. They are lined by a ciliated, mucus-secreting mucous membrane, and the mucus produced is carried towards the nasal cavity by the beating of the *cilia*. The *frontal* air sinuses are in the anterior part of the frontal bone above the orbital margin and the root of the nose. The paired *maxillary* sinuses are the largest, and lie above the back teeth. The *ethmoidal* air sinuses are small thin-walled cavities between the orbit and the upper part of the nasal cavity. The *sphenoidal* air sinuses in the body of the bone lie behind the ethmoid ones and above the cavities of the nose and nasopharynx.

THE LARYNX (*Figs 9b and c*)

This expanded upper portion of the *trachea* is modified to produce the voice as well as to act as an air passage. It lies below the hyoid bone and tongue, and consists of cartilages joined by ligaments, membranes and muscles and lined by mucous membrane. It is larger in the adult male than in the female. The *thyroid* cartilage, prominent in the male ('Adam's apple'), consists of two plates of cartilage joined at an angle in front and open behind. The *cricoid* cartilage is shaped like a signet ring.

The two pyramidal *arytenoid* cartilages sit side by side posteriorly on the upper edge of the cricoid lamina. The paired *corniculate* and *cuneiform* cartilages are small. The *epiglottis* is a leaf-like plate of elastic fibrocartilage; it helps to close the larynx and protect the lower air passages when food is swallowed.

The *vocal cords* are sharp white folds of mucous membrane, and the space between them is the *glottis*.

The larynx has extrinsic muscles passing between it and adjacent structures, and intrinsic muscles which lie entirely within it. The intrinsic muscles open and close the glottis and adjust the tension of the vocal cords, thus altering the pitch of the voice.

Arterial supply

From the laryngeal branches of the superior and inferior thyroid arteries.

Venous drainage

To the superior and inferior thyroid veins.

Lymphatic drainage

To deep cervical nodes.

Nerve supply

From branches of the vagus; sensory via the internal laryngeal nerve; the external laryngeal nerve is motor to the crico-thyroid muscle, while the recurrent laryngeal nerve supplies all the remaining intrinsic muscles.

THE TRACHEA (Windpipe) (*Fig. 10a*)

Half the trachea is in the neck and half within the thorax. It runs downwards from the larynx to divide into two main bronchi.

The trachea consists of incomplete rings of cartilage joined by fibrous tissue and unstriped muscle and lined with mucous membrane. There are 16–20 *tracheal rings*, which form arches round its front and sides. Posteriorly their tips are joined by fibromuscular tissue, forming a flat back wall. This construction keeps the airway open during neck movement, while allowing stretching during respiratory movements. The *respiratory epithelium* lining the trachea and the rest of the respiratory tract has mucus-secreting cells and others which are ciliated. The cilia beat upwards, so an ascending sticky mucous carpet collects and removes debris from the respiratory tree.

The last tracheal ring is thick and broad and has a process called the *carina* (keel) which runs backwards where the trachea branches into the right and left main *bronchi*. Outside the lungs, the bronchi are kept patent by incomplete rings like those of the trachea; within the lung substance they have irregularly distributed cartilaginous plates in their walls. Inside the fibrous tissue and cartilage is a layer of circularly arranged smooth muscle fibres, whose contraction reduces the diameter of the tube. It is lined by mucous membrane.

After entering the lung, the bronchi divide on the right side into three lobar bronchi and on the left into two, named according to the lobes of the lungs they supply. They divide again and again, forming smaller and smaller tubes until their terminal branches, the respiratory bronchi, divide into *alveolar ducts* and open into the *alveoli*. These are lined by a very thin flat epithelium and are surrounded by the pulmonary capillaries. Gas exchange takes place here, and the alveoli make up the substance of the lung.

THE LUNGS (*Fig. 10*)

These are two light, soft and spongy organs lying within the chest cavity on either side of the *mediastinum* (*Fig. 10b*). In newborn babies they are pink, but become greyish or blackened by carbon particles in adult town dwellers. Each is attached by its root at the *hilus* to the trachea and heart. The two lungs differ slightly from each other in that:

(1) The left lung is divided into two lobes by a long, deep oblique fissure, while the right lung is three-lobed, having both oblique and horizontal fissures.
(2) The right lung is slightly larger than the left.
(3) The anterior margin of the right lung is approximately straight, while that on the left is interrupted by the cardiac notch.

THE RESPIRATORY SYSTEM

The lung has a dual *blood supply*, from:

(1) the *pulmonary artery* (*Fig. 10c*), whose branches form the pulmonary capillaries on the alveolar walls. Capillaries leaving the alveoli join, eventually forming the *pulmonary veins*.
(2) the much smaller *bronchial arteries*, which arise from the descending thoracic aorta and the upper *intercostal arteries*, run with the bronchi and nourish the lung substance itself. The *deep bronchial veins* communicate with the pulmonary veins, while the *superficial* ones drain into the *azygos* and *superior intercostal veins*.

Lymphatic drainage

Both superficial and deep vessels drain into the nodes at the hilus or those around the trachea and large bronchi.

Nerve supply

From the vagus and the sympathetic system, via the anterior and posterior pulmonary plexuses.

THE PLEURA (*Fig. 10a*)

On each side of the body the pleura forms a closed sac invaginated by the lung like a fist pushed into a soft balloon. Each sac thus acquires two layers:

(1) A *visceral* or *pulmonary* layer which covers the lung surface and lines the fissures between the lobes.
(2) A *parietal* layer which lines the inner surface of the chest cavity.

The two layers are in contact during all phases of respiration and are lubricated by *pleural fluid*.

The Digestive System
(*Figs 11, 12, 13, 14 and 15*)

This consists of:

(1) the *alimentary canal* (gut), a muscular tube about 9 metres (30 ft) long, stretching from mouth to anus. It includes the:

  (*a*) mouth cavity
  (*b*) pharynx
  (*c*) oesophagus
  (*d*) stomach
  (*e*) small intestine
  (*f*) large intestine.

From the oesophagus onwards the gut tube has the same basic structure:

  (*a*) an innermost lining of mucous membrane
  (*b*) a submucous layer of connective tissue, carrying blood vessels and lymphatics
  (*c*) an inner layer of circular muscle, thickened in places to form sphincters
  (*d*) an outer layer of longitudinal muscle
  (*e*) an outermost coat of fibrous tissue (pharynx, oesophagus and rectum) or of serous membrane (rest of gut).

(2) *accessory organs*:

    (*a*) teeth
    (*b*) salivary glands (3 pairs)
    (*c*) liver and biliary system
    (*d*) pancreas.

THE MOUTH (*Fig. 11e*)

The mouth is the upper opening to the gut. It is surrounded by the lips, which consists of the *orbicularis oris* muscle, with vessels, nerves, labial glands and areolar tissue, covered with skin externally and mucous membrane internally.

The mouth cavity consists of the *vestibule*, the cleft between the lips and gums and teeth, and the *mouth cavity proper*. Within it are the teeth and the tongue; the *salivary glands* open into it.

The Teeth (*Fig. 11a, b and c*)

There are two sets (*Fig. 11a and b*)

    20 *milk* or *deciduous* teeth, erupting during the 1st and 2nd year:
    32 *permanent* teeth. These start to erupt and replace the first set during the 6th year.

The Tongue (*Fig. 11e*)

This is important in taste, speech and swallowing. It lies partly in the mouth cavity and partly in the pharynx and consists of muscle covered by mucous membrane. Its root is attached to the hyoid bone and the mandible. Its undersurface is covered with smooth mucous membrane, part of which forms a fold, the *frenulum*, connecting the tongue to the floor of the mouth. The anterior two-thirds of the upper surface of the tongue is covered with *papillae*. *Taste buds* are most numerous on the posterior third.

Arterial supply

Via the lingual arteries from the external carotids.

Venous drainage

Via the lingual veins to the internal jugular.

Lymphatic drainage

Via a plexus to the submental, submandibular and then the deep cervical nodes.

Nerve supply (see Fig. 23c)

Taste sensation is carried to the brain from the anterior two-thirds of the tongue by the *facial* nerve (VII) (*chorda tympani*), and from the posterior third by branches of the *glossopharyngeal* (IX) nerve. Common sensation is transmitted by the *lingual* nerve, from the mandibular branch of the *trigeminal* (V) nerve. The motor nerve for the muscles of the tongue is the *hypoglossal* (XII).

The Salivary Glands (*Fig. 11d*)

These are three pairs of large glands, the *parotid*, the *submandibular* and the *sublingual*, plus many small glands in the mucous membrane lining the mouth. The saliva they produce is the lubricant for speech and swallowing. It allows food to dissolve and reach the taste buds and starts the digestive process.

Parotid

The parotid glands are the largest and lie on the side of the face below and in front of the external auditory meatus. The parotid duct, about 5 cm long, runs forwards from the inner surface of the gland and opens into the mouth opposite the 2nd upper molar.

Submandibular

The submandibular glands, each about the size of a walnut, lie on the floor of the mouth just inside the mandible. Their ducts open beneath the tongue.

Sublingual

The sublingual glands, the smallest, are almond-shaped, and lie below the mucous membrane of the floor of the mouth behind the front teeth.

THE PHARYNX (*Fig. 15a*)

This is a funnel-shaped muscular tube lined by mucosa. It extends from the base of the skull to the level of the 6th cervical vertebra. It has three parts: nasal, oral and laryngeal.

Nasopharynx

The nasopharynx is a bony cleft lined with mucosa which lies behind the nasal cavity. The soft palate forms its floor and during swallowing lifts to separate it from the oropharynx. The *auditory (Eustachian) tubes* open into the lateral walls of the nasopharynx on each side. The *pharyngeal tonsils* ('adenoids') lie on its posterior wall.

Oropharynx

The oropharynx extends from the soft palate to the upper border of the epiglottis. Anteriorly is the opening from the mouth cavity and the root of the tongue, while behind lie the bodies of the 2nd and 3rd cervical vertebrae. On either side are the *palatopharyngeal* and *palatoglossal arches* (pillars of the fauces) between which lie the *palatine tonsils* ('tonsils').

The tonsils and adenoids form part of a ring of lymphoid tissue (*Waldeyer's ring*) around the entrance to the digestive and respiratory tubes. Lymph vessels pass from the tonsil to the upper deep cervical lymph nodes.

Laryngopharynx

The laryngeal part of the pharynx (laryngopharynx) runs downwards from the upper border of the epiglottis to end at its junction with the oseophagus opposite the cricoid cartilage. In front lie the cartilages and the opening of the larynx.

Arterial supply

From the external carotids via the facial, maxillary and lingual arteries.

Venous drainage

Via the pharyngeal plexus to the internal jugular.

Lymphatic drainage

To the deep cervical nodes.

Nerve supply

Mainly from the pharyngeal plexus (glossopharyngeal and vagus, with sympathetic). The main motor supply is from the vagus.

THE OESOPHAGUS (*Figs 14 and 15d*)

This is a muscular tube about 25 cm long running downwards from the pharynx, perforating the diaphragm, and ending at the cardiac opening of the stomach. Within its fibrous and muscular coats its mucous coat is thrown into longitudinal folds.

Arterial supply

From the inferior thyroid artery above, the thoracic aorta in the middle, and the left phrenic and left gastric arteries below.

Venous drainage

From the upper and thoracic parts into systemic vessels; and from the abdominal portion partly into the left gastric vein, a tributary of the portal vein and partly into the azygous, a systemic vein.

Lymphatic drainage

Into deep cervical nodes above, posterior mediastinal nodes in the middle, and the left gastric nodes below.

Nerve supply

From the vagus and the sympathetic.

THE STOMACH (*Figs 12 and 15d*)

This is continuous with the oesophagus above and the duodenum below and lies in the left hypochondrium and epigastrium. Within its outer serous coat of peritoneum and its layers of longitudinal and circular muscle is an oblique muscle coat. The arrangement of the fibres of the three muscle layers enables the stomach to churn, mix and break down the food within it. Glands in the folded mucous membrane contain cells which secrete the acid gastric juice and a protective mucus which shields the stomach lining itself from its action.

The stomach has two openings, two curvatures (greater and lesser) and two surfaces, anterior (superior) and posterior (inferior). The cardiac orifice or *cardia* is the upper opening from the oesophagus. The *pylorus* is the lower opening; the pyloric sphincter controls passage into the duodenum.

Arterial supply

From the coeliac axis (*Fig. 34a*), via the left and right gastric and right gastroepiploic branches of the common hepatic artery, and left gastroepiploic and short gastric branches of the splenic artery.

Venous drainage

Into the splenic or superior mesenteric veins (portal tributaries), or directly into the portal vein (*Fig. 38*).

Lymphatic drainage

To the lymph nodes lying along its curvatures and to the nearby hepatic nodes beside the hepatic artery in the lesser omentum.

Nerve supply

The sympathetic is from the *coeliac plexus*, and the parasympathetic via the *vagus* (*Fig. 24b*).

THE SMALL INTESTINE (*Fig. 12*)

This is about 6 metres long from the stomach to the *ileocaecal valve*, where it joins the large intestine. It has three parts: *duodenum*, *jejunum* and *ileum*.

The Duodenum (*Figs 13 and 15d*)

This is approximately 25 cm long and curves around the head of the pancreas. The bile duct and pancreatic duct enter it at the *ampulla of Vater*.

The rest of the small intestine consists of a greatly coiled mass of loops which occupy most of the abdominal cavity. It is covered in *peritoneum* which forms the fan-shaped fold of its *mesentery*, by which it is attached to the posterior abdominal wall. The mesentery contains blood vessels, nerves, lymph nodes and fat.

The intestinal glands found in the mucous membrane throughout the small intestine secrete intestinal juice which, with the pancreatic juice and helped by the emulsifying action of bile, completes the process of digestion.

The Jejunum

This is approximately 2 metres long. Its mucosa is thrown into many permanent circular folds which increase its surface area and slow down the passage of food. Thin-walled mucosal projections, villi, absorb the nutrients. They consist of a plexus of blood vessels and lacteals covered by mucosa.

The Ileum

This is 3½–4 metres long and opens into the large intestine at the junction of the caecum and the ascending colon. Rounded areas of lymphoid tissue (*Peyer's patches*) are found in the mucous membrane.

Arterial supply

From branches of the superior mesenteric artery (*Fig. 34b*).

Venous drainage

Into the superior mesenteric vein.

Lymphatic drainage

From lacteals and lymphoid tissue to the lymph nodes along the mesentery, and thence to the nodes on the front of the aorta.

Nerve supply

Sympathetic and parasympathetic via the coeliac and superior mesenteric plexuses.

THE LARGE INTESTINE (*Figs 12 and 13*)

This is wider than the small intestine but shorter, approximately 1.5 metres long, and frames the coils of the small intestine on three sides. It has the same basic structure as the rest of the gut, but longitudinal muscle bands (*taenia coli*) draw the bowel up into haustrations.

The large intestine is divided into the following:

(1) the *caecum*, into which the ileum opens. It lies in the right iliac fossa and has a blind rounded end. The ileal opening is protected by the ileocaecal valve which prevents backward flow;

(2) the *vermiform* (worm-like) *appendix*, a narrow, blind-ended tube containing lymphoid tissue which opens into the posteromedial surface of the caecum. Its length and position are variable;

(3) the *ascending colon*, approximately 15 cm long, which is narrower than the caecum. It runs upwards to the under surface of the liver, where it bends to form the

(4) *hepatic (right colic) flexure* and meets

(5) the *transverse colon*, approximately 50 cm long, which runs across the abdomen from the hepatic flexure to

(6) the *splenic (left colic) flexure*, where it becomes the descending colon. The splenic flexure lies under the left ribs against the lateral end of the spleen, the tail of the pancreas and the left kidney;

(7) the *descending colon*, approximately 25 cm long, which descends vertically as far as the iliac crest and then curves downwards and medially in an S-shape until it becomes the sigmoid colon at the pelvic inlet;

(8) the *sigmoid colon*, which forms a loop about 40 cm long within the pelvis, ending as

(9) the *rectum*, approximately 12 cm long, which curves downwards to become the anal canal. The upper part of the rectum has the same diameter as the sigmoid, but the lower part expands to form the rectal *ampulla*;

(10) the *anal canal*, approximately 3.8 cm long, which runs downwards and backwards to open to the outside world at the puckered *anus*. The *internal sphincter* of thickened circular muscle surrounds the upper three-quarters, while the *external sphincter* extends the whole length of the canal. The spincters keep the canal closed except during defaecation. The lowest 1 cm of the anal canal is lined by skin and the rest by mucous membrane.

Large Intestine above Anal Canal

Arterial supply

From the superior mesenteric artery as far as two-thirds of the way along the transverse colon (*Fig. 34b and c*); via inferior mesenteric artery branches from there as far as the upper half of the anal canal.

Venous drainage

Via the superior and inferior mesenteric veins which serve the areas supplied by the corresponding arteries (*Fig. 38*).

Lymphatic drainage

To nodes along the colic borders and mesenteries and thence by vessels running along the course of the arteries to the aortic nodes.

Nerve supply

Sympathetic from the abdominal ganglia and parasympathetic from the vagus, the fibres running with the arteries. The parasympathetic supply for the left half of the colon runs in the pelvic splanchnic nerves (nervi erigentes) (*Fig. 27*).

Anal Canal, Upper Part

Arterial supply

Via the superior rectal artery, from the inferior mesenteric.

Venous drainage

Via the superior rectal vein and inferior mesenteric vein to the portal vein.

Lymphatic drainage

With that of the rectum.

Nerve supply

From sympathetic nerves responsive to stretch but relatively insensitive to touch, temperature and pain.

Anal Canal, Lower Part

Arterial supply

Via the inferior rectal artery from the internal pudendal, a branch of the internal iliac artery.

Venous drainage

To the internal pudendal vein.

Lymphatic drainage

With the perianal skin to the superficial inguinal lymph nodes.

Nerve supply

The mucosa is supplied by spinal nerves, sensitive to pain.

THE LIVER (*Figs 12 and 13*)

This is the largest gland in the body. It is wedge-shaped, vascular and easily torn, and lies mainly in the upper right part of the abdominal cavity. It has a large right and a smaller left lobe. The *falciform ligament* separates the left and right lobes and connects the liver to the diaphragm and the anterior abdominal wall.

The *porta hepatis* (*Fig. 32b*) is a deep fissure on the undersurface of the liver where important structures enter and leave it. These are:

(1) the *portal vein* which carries blood from the gut, spleen, pancreas and gall bladder to the liver. Within its substance it branches repeatedly, eventually forming capillary-like vessels called *sinusoids*. These drain into hepatic veins which eventually pass their blood into the inferior vena cava.
(2) the *hepatic artery*, a branch of the coeliac artery, enters and supplies the liver tissue.
(3) the *hepatic plexus* of nerves, containing sympathetic and parasympathetic fibres, accompanying the hepatic artery.
(4) the right and left *hepatic ducts*, branches of the biliary tree, leave the liver at the porta hepatis, as do
(5) the *lymph vessels*, which carry a protein-rich lymph and end in the hepatic group of nodes. Other vessels run with the hepatic veins to the nodes around the inferior vena cava.

THE BILIARY SYSTEM (*Figs 12, 13 and 15d*)

Within the substance of the liver, tiny *bile canaliculi* lie between the cells and collect bile as it is produced. The canaliculi join together into larger vessels, eventually forming the right and left *hepatic ducts*. These leave the liver at the porta hepatis and join to form the common hepatic duct, which is about 3 cm long.

The Gall Bladder

The gall bladder (*Figs 12 and 32b*) lies beneath the liver. It has three coats: an outer, serous coat of peritoneum; a fibromuscular coat which contracts to expel bile; and a mucous coat, folded into a honeycomb arrangement which can absorb water and concentrate the bile. The cystic duct drains the gallbladder and joins the common hepatic duct on the right to form the bile duct, which descends to enter the duodenum at the ampulla of Vater. Around the lower part of the bile duct the circular muscle is thickened to form the *sphincter of Oddi*.

PANCREAS

The pancreas (*Figs 13 and 15d*) is a leaf-shaped creamy pink gland 12–15cm long lying across the posterior abdominal wall. Its rounded head, on the right side, lies within the curve of the duodenum and its tail against the spleen.

The *pancreatic duct* runs along the long axis of the gland from left to right, receiving ductules from the pancreatic lobules along its course in a herringbone fashion. In the neck of the gland it turns downwards, backwards and to the right to join the bile duct and enter the duodenum at the ampulla of Vater.

The pancreas is a dual purpose gland. Its *exocrine* secretion (via the duct) contains important digestive enzymes. The *endocrine* cells of the *islets of Langerhans* produce the hormones insulin and glucagon, important in glucose metabolism; these pass into the bloodstream.

THE PERITONEUM (*Figs 13, 15c, 15e and 18*)

The peritoneum, a large, serous membrane, has two layers: *parietal*, lining the abdominal cavity, and *visceral*, covering its contents, with a small quantity of lubricating fluid between them. The peritoneum of the *greater sac* lines the abdominal cavity, covers the

THE DIGESTIVE SYSTEM 67

diaphragm and is reflected down over the liver. It then runs downwards to the stomach, forming one layer of the *lesser omentum* that lies between the two. On the right side, its free border contains the hepatic artery, bile duct and portal vein, nerves and lymph nodes.

Descending from the stomach, the peritoneum forms the most anterior layer of a large apron-like free fold, the *greater omentum*. At the lower free margin of the greater omentum, the peritoneum turns upwards to reach the posterior abdominal wall. From here it extends downwards to reach and cover the transverse colon, pancreas and duodenum. It then descends to cover the jejunum and ileum. Returning to the posterior abdominal wall, it is reflected over the pelvic viscera, and then runs upwards inside the abdominal wall.

Folds of peritoneum enclosing major organs meet behind them to form a double sheet of membrane, the *mesentery*, which carries blood vessels, lymphatics and nerves, and is reflected back to secure them loosely to the posterior abdominal wall.

The *lesser sac* is a pocket in the greater sac forming a peritoneal recess behind the stomach. It extends upwards to the liver and downwards behind the stomach and within the greater omentum, contributing the inner two to its eventual four layers.

THE ABDOMINAL WALL (*Figs 7 and 12*)

From outside inwards this consists of:

(1) skin
(2) superficial fascia containing fat
(3) deep fascia
(4) superficial muscles, namely:

>(*a*) *external oblique*, running downwards and medially to join its fellow of the opposite side at the *linea alba*.
>(*b*) *internal oblique*, crossing the external oblique and fanning

out upwards and forwards from the inguinal ligament, the iliac crest and neighbouring fascia to the ribs.

(c) *transversus abdominis* running across the abdomen deep to the oblique muscles. It forms a central aponeurosis whose lower fibres join with the aponeurosis of the internal oblique to form the *conjoint tendon*, which is inserted into the pubis.

Immediately above the pubis the spermatic cord in the male or round ligament in the female (*Figs 13 and 17a*) pierces the external oblique. This hole is the *superficial inguinal ring*. The lower border of the external oblique, stretching from the anterior superior iliac spine to the pubic tubercle, forms the *inguinal ligament* (*Fig. 17a*).

(d) The *rectus abdominis* is a long, strap-shaped muscle running vertically down the abdomen from the costal margin to the pubis and meeting the rectus of the opposite side at the linea alba.

(5) The transversalis fascia lines the internal surface of the transversus abdominis muscle. It has an oval aperture, the *deep inguinal ring*, halfway along the course of the inguinal ligament and 1.25 cm above it.

(6) Internal to the transversalis fascia lie the extraperitoneal fat and the peritoneum.

THE INGUINAL CANAL (*Figs 13 and 17a*)

The inguinal canal is an oblique passage 6 cm long, slanting downwards and medially from the deep to the superficial inguinal ring. It contains the ilioinguinal nerve and the spermatic cord in the male, or the round ligament in the female.

The Urinary System
(Figs 14, 16, 17a and 18a)

This comprises the kidneys, ureters, bladder and urethra. The kidneys eliminate waste products and foreign substances from the blood and maintain the concentration of normal blood constituents such as salt and water. The urine produced by the kidneys drains down the ureters into the bladder, where it is stored until it is convenient to void it through the urethra.

THE KIDNEYS *(Fig. 16a and b)*

The kidneys are reddish-brown, soft and bean-shaped and lie behind the parietal peritoneum on the posterior wall of the abdomen. The left kidney is slightly higher than the right. The renal artery and nerves enter the organ and the vein and ureter leave it at the *hilus*. Within the kidney the *ureter* dilates into its *pelvis* and then into two or three major *calyces* (singular: calyx). Each of these subdivides into several minor calyces, which expand to surround one or more *renal papillae*.

The kidneys are embedded in perinephric fat and the whole is wrapped in fascia. Within the fat, the kidneys are enclosed by a fibrous capsule, which surrounds an outer *cortex* and inner *medulla*. The dark reddish-brown medulla consists of the conical *pyramids* whose apices form the renal papillae. The paler cortex surrounds the bases of the pyramids and dips down between their sides.

The kidney substance consists of a mass of uriniferous *tubules* with vessels, lymphatics, nerves and binding connective tissue. Each tubule consists of a *nephron* and a *collecting tubule*. The nephron forms the urine. It includes the *glomerular (Bowman's) capsule* in which ultrafiltration of plasma takes place, and the *renal tubule* within which by reabsorption and secretion the composition of the urine is tailored to the body's current requirements. The renal tubule has three parts:

(1) the *proximal convoluted tubule*, immediately below the glomerulus
(2) a thin hairpin bend, the *loop of Henle*
(3) the *distal convoluted tubule*.

This joins the *collecting tubule*, which doubles back, passing down into the medulla to open into the renal pelvis at the apex of the renal papilla.

Arterial supply

Each *glomerulus* contains a tangle of tiny blood vessels, the glomerular tuft. The afferent branch is derived from the renal artery, and the efferents branch to form a capillary network which runs between the tubules and both secretes into them and absorbs from them.

Venous drainage

These intertubular capillaries unite to form interlobular veins which eventully drain into the renal vein.

Lymphatic drainage

Via plexuses of vessels to the lateral aortic nodes.

Nerve supply

Branches of the coeliac and aortic plexuses form the renal plexus;

most nerves are vasomotor.

THE URETERS (*Figs 14 and 16a*)

Each ureter is a muscular tube beginning in the hilus of the kidney as the funnel-shaped pelvis of the ureter and narrowing to a thin tube which carries urine to the bladder. The ureter has three coats,

(1) *fibrous*, continuous with the renal capsule above and the outer coat of the bladder below,
(2) *muscular*, which propels the urine along the ureter, and
(3) longitudinally folded *mucous lining*, which is reflected over the renal papillae.

Each ureter runs downwards to the bladder and obliquely through its wall.

THE URINARY BLADDER (*Figs 17a and 18a*)

This is a muscular bag which stores urine until it can be voided, usually when it contains about 250 ml, but it can hold much more. When empty the bladder lies entirely within the pelvis, but as it fills it rises up into the abdominal cavity. The bladder neck is the lowest and most fixed portion, joining the urethra 3–4 cm behind the lower part of the pubic symphysis. The apex faces forwards and from it the *median umbilical ligament* runs upwards to the *umbilicus*.

The bladder wall consists of three coats:

(1) *serous*, of visceral peritoneum
(2) *muscular*, consisting of the criss-crossing unstriped fibres of the *detrusor muscle*
(3) *mucous*, continuous with the lining of the ureters above and of the urethra below. In the empty state, this lining is

thrown into folds, which allow it to stretch as the bladder fills. Its epithelium is of the transitional type, which also allows for distension.

Urine leaves the bladder through the internal urethral orifice. Here the bladder neck is guarded by the *internal sphincter* (*sphincter vesicae*), whose fibres are so arranged that the sphincter is opened when the detrusor fibres contract to expel urine. It is composed of involuntary muscle.

THE URETHRA (*Figs 17a and 18a*)

The urethra is 18–20 cm long in the male and approximately 4 cm long in the female, running from the bladder neck to the external urethral orifice or meatus at the penis or vulva. The *external sphincter* (*sphincter urethrae*) encircles it just above the perineal membrane.

The Male Reproductive System
(*Fig. 17*)

THE TESTES (*Fig. 17a*)

The testes hang suspended in the *scrotum*, the left usually lower than the right. They have a thick fibrous covering, the *tunica albuginea*. The *epididymis* is a coiled tube 6 metres long, folded into a comma shape and hooked above and behind the testis on the lateral side. At its lower pole the epididymis is continuous with the *vas deferens*, which has a thick muscular wall. The vas runs upwards and at the base of the *prostate gland* each is joined by the duct of the *seminal vesicle* to form the *ejaculatory duct*, a 2 cm long passsage which opens into the prostatic urethra. Muscular contractions of the seminal vesicles, prostate and vas deferens ejaculate the *semen* at orgasm. The semen consists of the sperm and the secretions of:

(1) the paired *seminal vesicles* which produce a creamy, alkaline fluid
(2) the *prostate gland*, whose thin, opalescent secretion is passed into the urethra by several small ducts. Shaped like a chestnut, it lies at the base of the bladder and around the commencement of the urethra. It consists of glandular tissue and muscle enclosed in a fibrous capsule. The urethra and the ejaculatory ducts both pass through it
(3) the paired, pea-sized *bulbo-urethral* (*Cowper's*) glands, lying on either side of the membranous urethra and passing their mucoid secretion into it.

THE SCROTUM (*Fig. 17a*)

The scrotum is a pouch containing the testes, epididymis and the lower part of the spermatic cords. It consists of:

(1) skin, loose, folded and covered in hairs and sebaceous glands
(2) the *dartos*, a thin sheet of unstriped muscle continuous with the scrotal septum, which divides the cavity into two halves
(3) the coverings of the spermatic cord and testis
(4) the parietal layer of the *tunica vaginalis*, whose two layers, separated by a film of fluid, allow the testis to slide within its coverings.

THE PENIS (*Fig. 17b*)

The root consists of three masses of erectile tissue: two *crura*, each attached to the *pubic rami* and converging and running forwards to become the two *corpora cavernosa* within the body, and the *bulb* of the penis, lying between the crura and running forwards to form the *corpus spongiosum* of the body, through which runs the penile (spongy) urethra. At its distal end, the corpus spongiosum enlarges to form the *glans penis*; the urethra opens on its tip. At its base, the glans has a projecting rim called the *corona* with a constriction, the neck of the penis, behind it.

The penile skin is loose at rest to allow for erection. The *prepuce* or foreskin partially covers the glans.

Arterial supply

Testis: via the testicular artery from the aorta. Penis: from the internal pudendal branch of the internal iliac artery via penile branches.

THE MALE REPRODUCTIVE SYSTEM

Venous drainage

Testis: by testicular veins to the pampiniform plexus and thence upwards to the inferior vena cava (via the renal vein on the left). Penis: *deep* to prostatic plexus and thence to the internal iliac vein; *superficial* into great saphenous vein.

Lymphatic drainage

Testis: to aortic nodes. Penis: to superficial inguinal nodes.

Nerve supply

Testis: via renal and aortic plexuses from the spinal 10th thoracic segment. Penis: from the 2nd, 3rd, and 4th sacral nerves (S2, S3 and S4), via the pudendal nerve and pelvic plexuses.

The Female Reproductive System
(Fig. 18)

THE OVARIES (*Fig. 18b*)

The ovaries are almond-shaped and are attached to the back of the broad ligament.

Arterial supply

Via the ovarian artery, from the abdominal aorta.

Venous drainage

Via the ovarian vein directly into the inferior vena cava on the right and on the left into the renal vein.

Lymphatic drainage

Into aortic nodes.

Nerve supply

From the ovarian plexus.

THE UTERINE (FALLOPIAN) TUBES

These lie in the upper margins of the broad ligament, are about 10 cm long and have fimbriated, funnel-shaped ends. They enter the uterus at the junction of the fundus and body. The tubes are muscular and their ciliated lining mucous membrane creates a current running towards the uterus.

THE UTERUS (*Fig. 18a and b*)

The uterus (womb) is a thick-walled, pear-shaped muscular organ, lying in the pelvis, with the bladder in front of it and the rectum behind. Its muscular wall is lined by a mucous coat, the *endometrium*. The rounded end of the uterus is the *fundus* and the main portion the *body*. This tapers downwards towards the internal os of the narrow *cervix (neck)*, which opens into the *vagina* at its external os. Nearly half the cervix lies within the vagina. The parts of the vagina encircling the cervix are the *fornices*.

Ligaments of the uterus (*Fig. 18a*)

These are:

(1) The *broad* ligaments, folds of peritoneum passing from the sides of the uterus to the sides of the pelvis, and containing the uterine tubes, vessels and nerves.
(2) The *anterior vesicouterine* ligament, running from the front of the uterus to the bladder.
(3) Two *uterosacral* ligaments, forming the lateral boundaries of the pouch of Douglas.
(4) The *transverse* (*cardinal* or *Mackenrodt's*) ligaments, strong fibromuscular structures running fanwise from the side of the cervix and upper vagina to the side of the pelvis, surrounding the blood vessels.
(5) The *round* ligaments, running from the lateral angles of the uterus through the inguinal canals together with blood vessels, nerves and lymphatics to end in the *labia majora*.

Arterial supply

From the uterine arteries (branches of the internal iliac) and the ovarian artery.

Venous drainage

By the uterine veins via plexuses into the internal iliac veins.

Lymphatic drainage

To adjacent nodes in the broad ligament, to nearby sacral and iliac nodes and along the round ligament to the superficial inguinal nodes.

The uterine tubes are supplied by ovarian and uterine vessels, and their lymph drains to the lumbar nodes.

Nerve supply

Autonomic from the hypogastric plexuses via the hypogastric nerve (sympathetic) and pelvic splanchnic nerves (parasympathetic).

THE VAGINA (*Fig. 18*)

The vagina runs downwards and forwards from the cervix to open onto the *vestibule* between the *labia minora*. The bladder and urethra are in front of it and behind from above downwards are the *rectouterine pouch (of Douglas)*, the rectum and the *perineal body*, which separates its opening from the anus. Its lining has transverse folds (*rugae*) which allow it to stretch.

Arterial supply

From the vaginal, uterine, internal pudendal and middle rectal branches of the internal iliac artery.

Venous drainage

Via plexuses to the internal iliac veins.

Lymphatic drainage

From the upper and middle part, to the external and internal iliac nodes. The lower part drains to the sacral, common iliac or superficial inguinal nodes.

Nerve supply

From the vaginal plexuses and from the pelvic splanchnic nerves. The lower part is supplied by the pudendal nerves.

FEMALE EXTERNAL GENITAL ORGANS (*Fig. 18c and d*)

The Vulva (*Fig. 18d*)

The vulva includes:

(1) the *mons pubis*, a pad of fat covered with skin and short curly hair overlying the symphysis pubis; from it,
(2) the two *labia majora* run backwards and unite behind, enclosing a cleft between them. They contain fat and are hairy outside; their inner surfaces are moist and smooth.
(3) the two *labia minora*, folds of skin lying within the labia majora on either side of the vaginal opening (*introitus*), which in the virgin is partially closed by the *hymen*. Bartholin's glands lie in the vestibule on either side of the vaginal opening and secrete a lubricant.
(4) the *clitoris*, lying in the anterior fold of the labia minora, and in front of
(5) the *urethral opening*, which lies just in front of the vagina.

The Endocrine System
(Fig. 19)

Endocrine glands produce *hormones*, which have a widespread action on body processes and pass directly into the bloodstream.

THE PITUITARY GLAND (hypophysis cerebri) *(Fig. 19)*

This is about the size of a large pea and is situated almost exactly at the centre of the head. It occupies the *pituitary (hypophyseal) fossa* of the sphenoid bone *(sella turcica)*. The fossa is roofed by a fold of dura called the *diaphragma sellae (Fig. 19)*, which separates the upper part of the gland from the optic chiasma.

The pituitary gland has two parts: anterior and posterior.

The Anterior Pituitary *(Fig. 19)*

The *anterior* pituitary is connected to the *hypothalamus* by a system of veins which carry *neurotransmitters* and *polypeptides* affecting the gland's function. The hormones produced by the anterior pituitary pass via capillaries to nearby venous sinuses and thus to the general circulation.

The Posterior Pituitary *(Fig. 19)*

The *posterior* pituitary has direct nervous connections with the hypothalamus and functions as an extension of it. The hormones are made in the hypothalamus and pass directly down the nerve fibres to be absorbed into the blood stream and circulate round the body.

THE THYROID GLAND (*Figs 12 and 19*)

This is a brownish-red, very vascular organ lying at the front of the base of the neck. It is H-shaped, consisting of *right* and *left lobes* joined by an *isthmus*. Its follicular cells produce *thyroxine* (T_4) and *triiodothyronine* (T_3) and its parafollicular C cells secrete *calcitonin*, important in calcium metabolism.

Arterial supply

Via the superior thyroid artery from the external carotid and via the inferior thyroid artery from the thyrocervical trunk.

Venous drainage

Via a plexus which drains to the internal jugular and the left brachiocephalic veins.

Lymphatic drainage

To the thoracic and right lymphatic ducts.

Nerve supply

From nearby thoracic ganglia.

THE PARATHYROID GLANDS (*Fig. 19*)

There are usually four of these yellowish-brown ovoid glands about 6mm × 3mm × 2mm in size. They are situated between the posterior borders of the lobes of the thyroid gland and its capsule. The glands secrete *parathormone*, which acts with calcitonin to regulate the level of calcium in the blood.

THE THYMUS (*Fig. 19*)

This lies in the chest behind the upper part of the sternum, extending upwards into the neck, and is pinkish-grey, lobulated and soft. It varies in size with age and usually atrophies after puberty. It is thought to have immunological functions.

SUPRARENAL (ADRENAL) GLANDS (*Figs 16a and 19*)

These triangular, paired yellowish organs, about 50mm × 30mm × 10mm in size, perch like cocked hats on the upper poles of the kidneys. Each has two parts, the *cortex* and the *medulla*.

The suprarenal cortex produces *glucocorticoids*, *mineralocorticoids* and *sex hormones*. The adrenal medulla produces *adrenaline*, important in the 'fight or flight' reaction.

Arterial supply

From the abdominal aorta and from the inferior phrenic and renal arteries.

Venous drainage

To the inferior vena cava on the right and to the renal vein on the left.

Lymphatic drainage

To the lateral aortic glands.

Nerve supply

Mostly sympathetic, from the coeliac plexus.

ENDOCRINE FUNCTIONS OF OTHER ORGANS (*Figs 13, 17a and 18a and b*)

The following organs which are described more fully elsewhere (pp. 66, 76 and 73) have endocrine functions.

The Pancreas

Hormones *insulin* and *glucagon* regulate blood sugar.

The Ovary

Hormones *oestrogen* and *progesterone* control the female reproductive system.

The Testis

The interstitial cells produce *testosterone*, which maintains motility and fertilizing power of sperm, increases size and strength of skeletal muscle and affects male secondary sexual characteristics.

The Nervous System
(*Figs 20 – 29*)

The nervous system is responsible for communication, both within the body itself and between the body and its surroundings. It consists of:

(1) nerves and their endings
(2) specialized receptors
(3) ganglia and plexuses
(4) the central nervous system (CNS), that is,
    the brain,
      in turn comprising
        (*a*) the cerebrum (2 cerebral hemispheres)
        (*b*) the brain stem
            midbrain
            pons
            medulla
        (*c*) cerebellum
    and the spinal cord
(5) the autonomic nervous system.

The functioning unit of the nervous system is the *neuron* (*nerve cell*) (Fig. 20a). This consists of a *cell body*, feathery *dendrites* which intercommunicate with those of other cells, and a long *axon* (nerve fibre). Nerve impulses are propagated along the axon electrochemically. Some axons are *myelinated*, that is, covered with a fatty bandage of *lipoprotein* applied by a spiralling *Schwann cell*. They appear white, and are indented at the *nodes of Ranvier*, where Schwann cell 'bandages' meet. Some axons (for example, in the autonomic nervous system and in the sensory nerves serving pain or smell) are unmyelinated.

In the CNS, the grey matter consists of masses of nerve cells; the white matter is formed of nerve fibres.

Many axons gathered together form nerves. *Efferent* nerves (E for exit) control the movement of muscles and gland secretion. *Afferent* nerves (A for approach) carry impulses from sensory organs or nerve endings to the CNS. Some nerves have only one function, either motor or sensory, but most are mixed. A *ganglion* is a gathering of nerve cells. A *nerve plexus* is a network of axons.

A *synapse* (*Fig. 20b*) is a point where one neuron meets another, but there is no anatomical continuity between the cells. A chemical transmitter is released across the gap to stimulate the next neuron. A similar neurochemical process operates between the axon and the structure it supplies, for example, a gland or muscle.

THE CENTRAL NERVOUS SYSTEM (*Figs 20, 21 and 22*)

The Brain

The brain (*Fig. 20c*) is covered by three membranes (*meninges*):

(1) The *dura mater* is thick, dense and inelastic to protect the brain. Its outer (*endosteal*) layer sends blood vessels and fibrous processes into the cranial bones. Its inner (*meningeal*) layer projects four folds inwards, which act as septa and divide up the cranial cavity. They are:

   (*a*) the *Falx cerebri*, lying between the two cerebral hemispheres and running from the crista galli to the
   (*b*) *Tentorium cerebelli*, a tented layer of membrane lying between the occipital lobes of the cerebrum above and the cerebellum below;
   (*c*) the *Falx cerebelli*, below the tent.

(*d*) the *Diaphragma sellae*, roofing the pituitary fossa and perforated by the pituitary stalk.

The outer, endosteal layer of the cerebral dura ceases at the margin of the *foramen magnum*. The inner, meningeal layer is continued downwards over the spinal cord as the spinal dura mater, enclosing the spinal cord and its inner membranes as far as the lower border of the second sacral vertebra, and ensleeving the roots of the spinal nerves.

(2) The *arachnoid mater* covers the brain and spinal cord and ensheaths the cranial and spinal nerves loosely until they leave the skull or vertebral canal. It extends downwards through the foramen magnum over the spinal cord and cauda equina, finally ending level with the lower border of the 2nd sacral vertebra.

(3) The *pia mater* is a soft, fragile membrane of many fine blood vessels held together by strands of tissue, and covering the brain and spinal cord very closely. It dips into every brain sulcus and is 'tucked in' to form the choroid plexuses of the ventricles. The spinal pia mater is thicker and less vascular. It covers the cord closely, dipping into the anterior fissure and sheathing the cranial and spinal nerves.

Between the arachnoid and pia mater is the *subarachnoid space*, which contains *cerebrospinal fluid* and the larger brain blood vessels.

Cerebrospinal Fluid (CSF) (Fig. 21)

The cerebrospinal fluid (CSF) is a clear liquid, about 130 ml in volume, surrounding the brain and spinal cord. It buffers the brain against injury and pressure changes and acts as an exchange medium between blood and brain. CSF is formed by diffusion and active secretion in the choroid plexuses, which are capillary loops projecting into the ventricles of the brain.

From the lateral ventricles the fluid flows downwards through the *foramen of Monro* to the 3rd ventricle, thence along the aqueduct to the 4th ventricle and out into the subarachnoid space through the *foramina of Luschka and Magendie*. CSF fills the cerebromedullary cisterns around the pons and medulla, and flows upwards over the pia enveloping the brain and downwards around the spinal cord and through its central canal.

Fingers of the arachnoid, with CSF within them, stick up into the venous blood in the sagittal sinus. These are called *arachnoid granulations*: they pass the CSF within them back into the venous blood.

The Cerebrum or Cerebral Hemispheres (Fig. 20c)

These consist of a layer of grey matter (*cortex*) covering the white matter, within which clumps of cells are embedded at the upper end of the brain stem. These are the *thalamus*, the *basal ganglia* (*Fig. 22*) and the *hypothalamus* (*Fig. 19*).

The cerebral cortex is folded and pleated to increase its area. The folds are *gyri* and the furrows between them *sulci*, and the whole resembles a half walnut. The cerebral surface thus becomes marked out into lobes, *frontal*, *parietal*, *occipital* and *temporal*, named according to the skull bones overlying them.

Each cerebral hemisphere mainly receives information from, and exerts motor control over, the opposite side of the body (*Fig. 22*). The two hemispheres are joined across the midline by fibres of the *corpus callosum*.

Voluntary movement originates in the precentral gyrus of the frontal lobe, the *motor cortex*, where each part of the body is represented. From the *motor cortex*, fibres pass downwards through the brain as the *corticospinal tracts*. Most of the fibres cross the midline in the *decussation of the pyramids*, then running down the spinal cord as the *lateral corticospinal tract*. The uncrossed fibres form the *anterior corticospinal tract*, whose fibres cross in the white commissure of the spinal cord. The fibres of the corticospinal tracts, which synapse either with an anterior horn cell of the spinal cord or with the motor nucleus of a cranial nerve, eventually innervate the muscles. The fibres beyond the synapse are referred to collectively as the *lower motor neuron*, while those proximal to the synapse are together called the *upper motor neuron*.

The *temporal cortex* is concerned with taste, smell and hearing, and is also important in memory and emotion.

The *frontal lobes* help initiate complex movement and also with higher cortical functions such as planning and foresight and the formation and maintenance of psychological set.

The *basal nuclei (ganglia)*, the organizing centres of the *extrapyramidal system*, are masses of grey matter within the white matter of the cerebral hemispheres. They are:

(1) The corpus striatum
    (*a*) lentiform nucleus
        putamen (dark, lateral)
        globus pallidus (pale, medial)
    (*b*) caudate nucleus
(2) amygdaloid body
(3) claustrum

The Cerebellum

The *cerebellum*, lying below the cerebral hemispheres, consists of a pair of *cerebellar hemispheres* and a central *vermis* (worm), formed of grey matter covering white matter and partially divided by the falx cerebelli.

The basal ganglia and the cerebellum help control body movement. Tne vestibular apparatus of the inner ear is also connected because of its function in maintaining posture and balance (*Fig. 22*).

The Brain Stem

The *pons* (*Figs 21 and 22*) contains fibres linking the different parts of the brain with each other. Some cranial nerve nuclei also lie in it. It is about 3.75 cm long and 5 cm wide.

THE NERVOUS SYSTEM

The *medulla* (*Fig. 20c*) is continuous with the pons above and the spinal cord below. It is bulb-shaped and is about 3 cm long and 2 cm wide at its widest point. The fibres of its white matter are:

(1) the descending pyramidal tracts (motor), lying anteriorly and crossing the midline as the decussation of the pyramids in the low medulla;
(2) lying posteriorly, sensory fibres carrying information upwards to the brain.

The lower cranial nerve nuclei lie in the medulla: it also contains the vital centres which control respiration, temperature etc.

The medulla, pons and cerebellum together are known as the *hindbrain*.

THE CRANIAL NERVES (*Figs 24a and 35b*)

There are twelve named pairs numbered from above downwards depending on their position in the brain stem.

The Olfactory Nerves (I)

These are purely sensory and are concerned with smell. 'Scent' is transmitted from receptors in the nasal mucosa to the olfactory bulbs. The olfactory nerves carry it onward to the olfactory cortex in the temporal lobe, where it is analysed.

The Optic Nerves (II)

These (*Fig. 28b*) are also purely sensory. On each side, fibres from the nerve endings in the retina are collected into bundles as they leave the eyeball through the *optic disc* to form the optic nerve. After entering the cranial cavity each *optic nerve* runs backwards and medially to meet and join its fellow at the optic chiasma. Here fibres from the medial (nasal) sides of the retina cross. The fibres then run on in the optic tract to:

(1) the *lateral geniculate body*, and thence to the occipital (visual) cortex, where visual information is analysed;
(2) to the *III nerve nucleus*, for the pupillary light reflex;
(3) to the *superior colliculus* (*corpus quadrigeminum*), the reflex centre for co-ordinated movement of the head, eyes, trunk and limbs in response to visual stimuli.

The Oculomotor (III), Trochlear (IV) and Abducens (VI) Nerves (Figs 24a and 28c)

These supply the muscles that move the eyeball. The IV nerve innervates the *superior oblique* muscle and the VI the *lateral rectus*. The III nerve supplies the remaining muscles and also the *levator palpebrae superioris*, which raises the upper eyelid. It also carries parasympathetic fibres towards the pupil.

Running from their nuclei in the brainstem, the nerves traverse the cavernous sinus to enter the orbit through the superior orbital fissure (*Fig. 9a*).

The Trigeminal Nerves (V) (Fig. 23a)

These are mixed nerves, being:

(1) *sensory* to the face, the scalp as far back as the vertex, and to the mucous membranes of the paranasal sinuses, nose and mouth. They also supply common sensation to the tongue and the teeth.
(2) *motor* to the muscles of mastication (masseters, temporals, lateral and medial pterygoids).
(3) *secretomotor*, carrying parasympathetic fibres on part of their journey to the lacrimal gland.

The nuclei of the trigeminal nerves lie in the mid-pons. Each sensory root is connected to the trigeminal (semilunar) ganglion, lying in a depression at the apex of the petrous temporal bone. It branches from the ganglion into three divisions:

(1) ophthalmic nerve
(2) maxillary nerve
(3) mandibular nerve

The motor root runs directly from the pons to join the mandibular nerve.

The *ophthalmic* nerve runs forwards from the ganglion to enter the orbit through the superior orbital fissure. Its main branches are:

(1) the *lacrimal* nerve, carrying secretomotor fibres to the lacrimal gland, and eventually supplying the upper eyelid;
(2) the *frontal* nerve which, after leaving the orbit, supplies the forehead and scalp;
(3) the *nasociliary* nerve, which runs medially in the orbit where it gives off the long ciliary nerves that carry sympathetic fibres to the dilator pupillae. It descends to supply the nasal mucous membrane and the skin of the nose tip.

The *maxillary* nerve arises from the semilunar ganglion and runs forward through the cavernous sinus and the foramen rotundum. Its infraorbital branch enters the orbit and emerges onto the face through the infraorbital foramen. Alveolar branches from the maxillary and infraorbital nerves innervate the upper teeth and gums.

The maxillary nerve and its branches carry sensation from the paranasal air sinuses, mouth, palate, pharynx, tonsils, upper teeth and gums and the skin of part of the face.

The *mandibular* nerve is joined by the motor root and leaves the skull by the foramen ovale as a mixed nerve. It runs downwards in front of the neck of the mandible, and branches downwards and forwards through the face, supplying structures on the way. Its anterior division gives off:

(1) motor branches to the masseters, temporals and the lateral pterygoid muscle;

(2) sensory branches to part of the cheek and to the mucous membrane of the buccal cavity.

Its posterior division gives off:

(1) the *auriculotemporal* nerve, supplying sensation to part of the skin of the pinna of the ear, the external ear canal and drum, the temporomandibular (jaw) joint and the skin of the temple.
(2) the *lingual* nerve, which carries common sensation from the anterior two-thirds of the tongue, the floor of the mouth and the gums. It is joined by the chorda tympani branch of the facial (VII) nerve; these fibres carry taste sensation fibres from the anterior two-thirds of the tongue and are secretomotor to the submandibular and sublingual glands.
(3) the *inferior alveolar* nerve, which is sensory to the teeth and gums of the lower jaw. The mental branch carries common sensation from the chin and lower lip.

The Facial Nerves (VII) (Fig. 23b)

These are mixed nerves, being motor to the muscles of facial expression and sensory (taste) from the anterior two-thirds of the tongue (*Fig. 23c*).

Motor and sensory roots emerge at the lower border of the pons and pass laterally and forwards with the VIII nerve to enter the temporal bone through the internal auditory (acoustic) meatus. The nerve then runs through the facial canal, within which it forms the geniculate ganglion and gives off:

(1) the *chorda tympani*, which joins the Vth nerve,
(2) the *nerve to the stapedius muscle* of the middle ear,
(3) the *greater superficial petrosal* nerve, carrying taste fibres from the palate and also secretomotor (parasympathetic) fibres towards the lacrimal gland.

On leaving the skull, the nerve runs forward through the parotid gland and supplies all the muscles of facial expression.

The Vestibulocochlear (auditory) Nerves (VIII) *(Figs 24d and 29a)*

These have two parts, which leave the brain stem together and run forwards to enter the internal acoustic meatus of the petrous temporal bone. Each then divides into:

(1) the cochlear part, which carries auditory information from the *cochlea* to the auditory nucleus in the floor of the 4th ventricle. The fibres then run up the brain stem to the *medial geniculate body* and on to the auditory cortex in the temporal lobe, where sound information is analysed;
(2) the vestibular part, which supplies the sensory receptors in the three *semicircular canals*, the *utricle* and the *saccule*. These are sensitive to the position of the head and its movements in space. This information is transmitted along the nerve to the vestibular nuclei in the pons and medulla. These nuclei have connections with the centres controlling body movement, for example the motor nuclei and the cerebellum.

The Glossopharyngeal Nerves (IX) *(Fig. 24a)*

These are mixed nerves and supply

(1) motor fibres to the stylopharyngeus,
(2) common sensation to the nasopharynx and the posterior aspects of the soft palate and tongue,
(3) taste fibres to the posterior third of the tongue.

The nerves also carry parasympathetic (secretomotor) fibres for part of their course to the parotid gland.

The nuclei of the glossopharyngeal nerves lie in the medulla. Each IXth nerve leaves the skull by bending downwards through the jugular foramen, where it gives off branches to carry sensation from the ear and the mastoid ear cells, and also the secretomotor fibres for the parotid gland. It then supplies the stylopharyngeus muscle and enters the pharynx to reach the mucous membrane, the tonsil and the posterior part of the tongue.

The Vagus Nerves (X) (Fig. 24b)

These mixed nerves pass from their nuclei in the medulla, out of the skull through the jugular foramen and down through the neck and thorax to the abdomen. They are:

(1) motor to the voluntary muscles of the larynx and pharynx;
(2) sensory to the thoracic and upper abdominal viscera and to a small area in the external ear canal;
(3) carry parasympathetic innervation to the viscera.

The Accessory Nerves (XI) (Fig. 24a)

These motor nerves each have two parts. The *cranial* portion is really only an accessory to the vagus and arises in the nucleus ambiguus in the medulla. It joins the *spinal* portion as it passes through the jugular foramen and separates from it again at its exit, to join the vagus.

The *spinal* root arises from the upper cervical segments of the spinal cord and runs upwards to enter the skull through the foramen magnum. Crossing to the jugular foramen, it descends through it, connecting with the cranial portion en route. It descends through the neck to supply the sternomastoid muscle and the trapezius.

The Hypoglossal Nerves (XII) (Fig. 24a)

These are the motor nerves to the muscles of the tongue. From the medullary nuclei the fibres run from the brain stem to the hypoglossal canal (anterior condylar canal) in the occipital bone. Emerging deep in the structures of the neck, each nerve descends to the level of the angle of the jaw where it turns and runs forward to supply the muscles of the tongue. It communicates freely with the lingual nerve.

The Spinal Cord (*Figs 20, 21, 22 and 27*)

This extends downwards from the medulla as far as the *cauda equina* (horse's tail). Its core, of grey matter, consists largely of nerve cells, is roughly H-shaped in cross-section and is surrounded by columns of white matter, formed of ascending and descending tracts of nerve fibres. The posterior, dorsal half of the cord serves sensory activity (ascending tracts) while the anterior (ventral) half is motor (descending tracts). The grey matter has right and left, anterior (ventral) and posterior (dorsal) horns.

The spinal cord lies in the spinal canal, surrounded by meninges and by CSF in the subarachnoid space (see above, p. 86). The central canal is also filled with CSF.

Along the cord, posterior horns receive sensory fibres and anterior horns give off motor ones. The fibres entering and leaving at the same level join up to form mixed spinal nerves, which emerge from the spinal canal via the intervertebral foramina.

Two posterior spinal arteries and an anterior spinal artery (*Fig. 35c*), branches of vertebral arteries, run along the length of the cord and branch to ring and supply it.

THE SPINAL NERVES (*Fig. 22*)

There are 31 pairs:

> 8 cervical
> The first spinal nerve (C1) arises above the atlas, and is sometimes called the *suboccipital* nerve; C8 (8th cervical nerve) arises below the 7th cervical vertebra
> 12 thoracic
> 5 lumbar
> 5 sacral
> 1 coccygeal

Because the cord is shorter than the canal, the nerves slant downwards to reach the intervertebral foramina. Below the cord they form the cauda equina. The tapering end of the cord, the *conus medullaris*, ends in the *filum terminale* lying within the cauda equina and attached to the back of the coccyx.

The *dorsal* (posterior) sensory root and the *ventral* (anterior) motor root perforate the dura separately before uniting into a spinal nerve. The *spinal ganglia* lie on the dorsal roots, usually in the intervertebral foramina through which the nerves leave the spinal canal.

After emerging from the intervertebral foramina, the spinal nerves split into dorsal and ventral rami, each of which carries both motor and sensory fibres. The dorsal rami run backwards to supply the muscles and skin of the back of the neck and trunk.

The ventral rami supply the muscles at the front and side of the trunk and, more importantly, the limbs. In all regions except the thoracic, these rami form plexuses. From above downwards these are: cervical, brachial, lumbar, sacral and coccygeal.

The cervical plexus

This is formed by the ventral rami of the upper 4 cervical nerves (C1, C2, C3 and C4).

Branches of the cervical plexus:

(1) supply motor innervation to the anterior and lateral muscles of the neck;
(2) supply sensation to the skin overlying these muscles, and to that of the side of the head;
(3) communicate with cranial nerves X, XI, and XII and with the sympathetic trunk;
(4) supply the diaphragm via the *phrenic nerve*, the most important branch of the plexus. This nerve descends to the root of the neck behind the sternomastoid and passes down through the thorax, where it supplies sensation to the mediastinal pleura, to the pericardium and to the parietal pleura and the peritoneum above and below the central tendon of the

diaphragm. It then branches to supply motor fibres to the muscle of the diaphragm.

The brachial plexus

This (*Fig. 25a*) forms from:

(1) the ventral rami of the lower 4 cervical nerves (C5, C6, C7 and C8);
(2) most of the ventral ramus of the 1st thoracic nerve (T1);
(3) small contributions from C4 and T2.

The plexus lies at the lateral border of the *scalenus anterior* muscle, behind the sternomastoid. The *roots* join to form *trunks*, which pass down through the neck to the clavicle, where each trunk splits into *anterior* and *posterior divisions*. These descend into the axilla around the axillary artery, splitting and reuniting into *lateral*, *posterior* and *medial cords*, which break up into the large *nerves* of the upper limb.

Branches of the brachial plexus supply the muscles of the shoulder and upper limb. Important branches include:

(1) *supraclavicular*:
  (a) the *long thoracic* nerve supplying the serratus anterior muscle;
  (b) the *suprascapular* nerve supplying motor innervation to the supraspinatus and infraspinatus and some sensation to the shoulder (glenohumeral) and acromioclavicular joints.

(2) *infraclavicular*:

  (a) the *pectoral* nerves supplying pectoralis major and minor;
  (b) the *subscapular* and *axillary* (*circumflex humeral*) nerves, supplying shoulder muscles;

(c) the *thoracodorsal* nerve to the latissimus dorsi.

(3) *Nerves to the muscles of the arm*:

(a) the *musculocutaneous* nerve supplies the biceps, the brachialis and the coracobrachialis;
(b) the *median* nerve runs downwards into the arm until, below the elbow joint, it supplies all the flexor muscles of the front of the forearm except the two supplied by the ulnar nerve. It then enters the hand and supplies the following short muscles:

lateral two lumbricals	L	
opponens pollicis	O	(mnemonic)
abductor pollicis brevis	A	
flexor pollicis brevis	F	

(c) The *ulnar* nerve runs down the arm, and at the elbow lies in a groove on the back of the medial epicondyle of the humerus. It enters the forearm on the ulnar side and supplies the two muscles of the front of the forearm not supplied by the median nerve:

flexor carpi ulnaris
part of flexor digitorum profundus.

It then runs downwards to supply all the small muscles of the hand apart from the 4 supplied by the median nerve (see above);

(d) the *radial* nerve is the largest branch of the brachial plexus. It descends behind the brachial artery (the continuation of the axillary artery) into the posterior part of the arm, where it supplies the triceps and anconeus and spirals round the humerus to the lateral side, giving off the posterior interosseous nerve in front of the lateral epicondyle. It then descends along the lateral (radial) side of the forearm, supplying the brachialis, the brachioradialis and the extensor carpi radialis longus. It ends by winding round the lateral side of the radius, becoming superficial and supplying cutaneous branches to the back of the first 3 digits on the lateral side of the hand.

(e) The *posterior interosseous* nerve winds round to the back of the forearm to innervate the supinator, the extensors of the wrists and the long extensors of the digits. It ends by supplying sensation to the wrist joint.

The ventral rami of the thoracic nerves do not form a plexus, but remain segmental as the *intercostal* nerves supplying the intercostal and abdominal muscles and the skin overlying them. The dorsal rami supply the muscles and skin of the back.

The lumbar plexus

This (*Fig. 26a*) is formed from:

(1) the ventral rami of the first 4 lumbar nerves, L1, L2, L3 and L4;
(2) a small contribution from T12.

It lies in front of the transverse processes of the lumbar vertebrae, in the substance of the psoas major muscle.

The arrangement of the plexus is somewhat variable.

Cutaneous branches of the nerves arising from the plexus supply the skin of the genitalia and that of the front and side of the thigh. Muscular branches supply the anterior and medial thigh muscles. They include:

(1) the *iliohypogastric* nerve which supplies twigs to the abdominal muscles and the skin of the side of the buttock and the lower part of the abdomen;
(2) the *ilioinguinal* nerve, which gives off small branches to the abdominal muscles and then runs through the inguinal canal

to supply the skin over the mons veneris and labium majus or of the scrotum and root of the penis, and the skin of the adjoining part of the thigh;
(3) the small *genitofemoral* nerve, supplying the external genitalia and the skin over the anterior part of the thigh;
(4) the *lateral cutaneous nerve of thigh*, which supplies the skin of the anterior and lateral parts of the thigh;
(5) the *obturator* nerve, which leaves the pelvis through the obturator foramen and supplies the adductor muscles to the thigh, the skin overlying them and a small branch to the knee joint;
(6) the *femoral* nerve (L2, L3, L4), the largest branch of the lumbar plexus; it runs down behind the inguinal ligament into the thigh, where it gives off cutaneous branches to the skin over the lateral and anterior surfaces. It also supplies the anterior thigh muscles (sartorius and quadriceps femoris) and gives off the *saphenous* nerve, which supplies the skin over the patella, along the medial side of the leg below the knee and of the medial side of the foot.

The Sacral Plexus

This (*Fig. 26a*) lies on the posterior wall of the pelvic cavity and is formed from:

(1) the *lumbosacral trunk* (L4, L5) which descends over the pelvic brim in front of the sacroiliac joint to join the 1st sacral nerve;
(2) the *ventral rami* of the first 3 sacral nerves and part of the 4th, that is, S1, S2, S3 (S4).

Its branches include:

(1) the *superior* and *inferior gluteal* nerves, which supply the muscles of the buttock;
(2) the *pudendal* nerve, which supplies sensation to the rectum, the perineum and the external genitalia, including part of

the vagina;
(3) the *posterior cutaneous nerve of the thigh*, which supplies the skin of the back of the thigh and of parts of the buttock and of the external genitalia;
(4) the *sciatic* nerve, which is the biggest nerve in the body. It leaves the pelvis through the greater sciatic foramen and descends in a curve through the buttock and the back of the thigh, dividing in the lower third into the *tibial (medial popliteal)* and *common peroneal (lateral popliteal)* nerves. During its descent it supplies the hip joint and the muscles of the back of the thigh.

The *tibial* nerve runs down through the back of the thigh and the popliteal fossa into the leg, where it supplies the calf muscles and the long flexors of the toes. At the ankle, it runs behind the medial malleolus into the sole of the foot where it supplies the small muscles, the overlying skin and the joints as the *medial* and *lateral plantar* nerves.

The *common peroneal* nerve, smaller than the tibial nerve, runs down the thigh and the popliteal fossa to wind round the neck of the fibula, where it divides to supply the muscles of the leg and foot.

The *sural* nerve, arising from both tibial and common peroneal nerves, supplies the skin of the lower calf and the lateral side of the foot and little toe.

(5) Through the *pelvic splanchnic* nerves (S_2, S_3, S_4), autonomic fibres are conveyed to the lower colon, the rectum, the bladder and the reproductive organs. These nerves are important in sexual intercourse and for the normal functioning of the bowel and bladder.

Muscular branches from S_4 run to the external anal sphincter and the levator ani.

The coccygeal plexus

This is formed from the remaining sacral and coccygeal nerves and supplies the skin over the coccyx.

THE AUTONOMIC NERVOUS SYSTEM (ANS) (*Fig. 27*)

The autonomic nervous system controls and adjusts those functions of the body that are mainly unconscious. It has close connections with the endocrine glands. The ANS supplies:

(1) smooth muscle (in bronchioles, gut, blood vessels, uterus and bladder)
(2) cardiac muscle
(3) glands

Special nerves run to the viscera, and the whole of the somatic peripheral nervous system carries autonomic fibres to the blood vessels and skin. Autonomic afferents transmit sensation, usually unconscious, from the viscera to the CNS. Sensations of hunger, nausea, distension of rectum or bladder and visceral pain are also probably transmitted in this way.

The autonomic nervous system comprises the *sympathetic* and *parasympathetic* systems.

Sympathetic (*Fig. 27*)

The sympathetic nervous system consists of the sympathetic trunks with their branches, plexuses and other ganglia.

The *sympathetic trunks* consist of two chains of ganglia on either side of the vertebral column, which meet in front of the coccyx in a terminal ganglion. There are usually 3 cervical, 11 thoracic, 4 lumbar and 4 or 5 sacral ganglia.

From the *superior cervical* ganglion the internal carotid nerve runs upwards, forming a plexus around the internal carotid artery. Branches from this supply blood vessels, glands and other structures in the head and neck.

Fibres from the *cervical* and *thoracic* ganglia run to the *sympathetic plexuses* round the viscera.

The *coeliac* (*solar*) plexus lies around the coeliac axis. Its branches supply the abdominal viscera and form secondary plexuses round the great vessels.

The *superior hypogastric* plexus (*presacral nerve*) lies between the common iliac arteries in front of the last lumbar vertebra and the sacral promontory. Its branches eventually supply the reproductive and excretory organs in the pelvis.

The functions of the sympathetic prepare the body for emergency actions in: fright, flight or fight. The whole body acts at once, normally in an upright position.

Connections with the central nervous system

The sympathetic nervous system is connected to the CNS in the thoracolumbar region of the spinal cord. Small myelinated fibres (*white rami communicantes*) run from the cell bodies in the lateral columns of the cord through the ventral root of the spinal nerve to the ganglia of the sympathetic trunk. Postganglionic, unmyelinated fibres run from each ganglion to the corresponding spinal nerves. Each of these carries fibres destined for the blood vessels, sweat glands and errectores pilorum muscles in its area. Other postganglionic fibres are distributed directly to visceral plexuses. Some fibres run up or down along the sympathetic trunk before leaving it.

Parasympathetic (*Fig. 27*)

This consists of fibres connected with the CNS in the cranial and sacral regions (craniosacral outflow). They usually synapse in ganglia near the organs they supply, and the system is therefore capable of a localized response.

Fibres arising in the midbrain join the IIIrd (oculomotor) cranial nerve and are relayed to supply the ciliary muscle and the sphincter pupillae. They contract the pupil and accommodate the lens for near vision.

Fibres arising in the medulla·

(1) travel in the facial, mandibular and glossopharyngeal nerves to reach and stimulate the secretion of the salivary glands;
(2) pass in the vagus to the thoracic and abdominal viscera.

Preganglionic parasympathetic fibres also run in the sacral nerves; they stimulate emptying of the lower bowel and supply the reproductive tract, including vasomotor fibres via the nervi erigentes to the erectile tissue of the penis or clitoris.

The activities controlled by the parasympathetic are those of sex, birth, digestion and emptying of the bowel and bladder. They usually happen with the body in a sitting or lying position.

The Special Senses
(*Figs 28 and 29*)

THE EYE (*Fig. 28*)

The eyeball is an almost spherical hollow structure whose wall has three layers: sclera, choroid and retina.

The Sclera (*Fig. 28a*)

The sclera, the outer coat, is tough and fibrous and helps the eyeball to keep its shape. Its posterior five-sixths is white and opaque, while anteriorly it is continuous with the transparent, avascular *cornea*. The tendons of the extrinsic eye muscles are inserted into the sclera.

The choroid (*Fig. 28a*)

The choroid, the middle coat, consists of interlacing blood vessels and pigment granules supported by connective tissue. Anteriorly, its outer layers form the *iris*, a coloured muscular diaphragm which controls the size of the *pupil*. The inner layers of the choroid form the *ciliary body*, which is pleated like an Elizabethan ruff into 60–80 ciliary processes, overlying a ring of ciliary muscle. The *lens* of the eye lies behind the iris, enclosed in an elastic capsule from which run suspensory ligaments, attaching it to the ciliary processes.

The anterior chamber between the cornea and iris, and the posterior chamber between the iris and lens, contain *aqueous humour*, which nourishes the cornea and lens. Aqueous humour is absorbed back into the circulation through the *canal of Schlemm*, running between the iris and cornea.

The retina (*Fig. 28a*)

The retina is the inner coat of the eye. Its outer pigment layer lies against the choroid, and the inner nervous layer has a light-sensitive part composed of *rods* and *cones*. The *optic nerve* leaves the eyeball at the *optic disc* (*blind spot*), carrying light sensation to the brain for analysis.

The Eyeballs and Eyelids (*Fig. 28a*)

Behind the lens and occupying four-fifths of the eyeball is the *vitreous humour* or *body*, condensed peripherally to form the hyaline membrane and in front into the radially arranged fibres of the suspensory ligament of the lens.

The eyeballs are moved synchronously within the orbit by the action of the extrinsic eye muscles, producing binocular vision (*Fig. 28c*).

The eyeballs lie in the bony cavity of the orbits, protected by the eyelids. These are lined by the thin *conjunctiva* which is reflected over the anterior surface of the eyeball and thus forms the closed conjunctival sac. The edges of the lids are fringed by eyelashes with related *sebaceous glands*, and studded by *tarsal (Meibomian) glands* whose oily secretion helps to stop overflow of the tears.

The eyelids are closed by the *orbicularis oculi* muscles, whose fibres lie within it and are supplied by the facial (VII) nerve. They are opened by the *levator palpebrae superioris* muscle; this is innervated by the III (oculomotor) nerve, which carries both somatic and sympathetic fibres. The slit between the eyelids is the *palpebral fissure*. The angle at the medial end is the *inner canthus*, and the *outer canthus* is the lateral angle.

The Lacrimal Apparatus (Fig. 5a)

The lacrimal apparatus produces and circulates the tears. These keep the conjunctiva and cornea moist, lubricated and nourished, and contain a bactericide, *lysozyme*.

The *lacrimal gland*, the size and shape of an almond, lies in the lacrimal fossa of the frontal bone and has about twelve *lacrimal ducts*. The contraction of the orbicularis oculi muscle during blinking milks the tears towards the *lacrimal canals*, whose openings are the tiny *lacrimal puncta* visible near the inner canthus. The canals drain into the *nasolacrimal duct*, and thence into the nose below the inferior concha.

THE EAR (*Fig. 29*)

This converts sound waves into a form which the brain can assimilate and analyse. It has three parts: the outer, middle and inner ear.

The Outer (external) Ear (*Fig. 29a*)

This consists of:

(1) The *auricle* or *pinna*, a curly leaf of cartilage covered by skin, with a soft, fatty and fibrous earlobe below.
(2) The *external acoustic* (*auditory*) *meatus* or *canal*. Its outer third is of cartilage, continuous with that of the pinna; the wall of the inner two-thirds is part of the temporal bone. The whole is lined with skin containing hairs and wax-secreting ceruminous glands.

At the junction of the outer and middle ear is the *eardrum* or *tympanic membrane* (*Fig. 29b*), about 9 mm across, lying obliquely and consisting of a central fibrous layer covered by skin on its outer surface and mucous membrane on the inner side.

The Middle Ear (*Fig. 29a*)

The middle ear or *tympanic cavity* is an air-filled space in the temporal bone. It is lined by mucous membrane and contains the three tiny *ossicles* which carry sound vibrations from the outer to the inner ear. The outermost bone, the *malleus (hammer)*, is attached by its handle to the eardrum, and a facet on its head forms a minute synovial joint with the *incus (anvil)*. This in turn articulates with the *stapes (stirrup)*. Two tiny muscles, the *tensor tympani* and the *stapedius*, damp down the system against the large vibrations caused by loud sounds. The footplate of the stapes is attached by the *annular ligament* to the rim of the *oval window*. The *auditory (Eustachian) tube*, about 3.5 cm long, connects the middle ear with the pharynx. Normally closed, during swallowing it opens to equalize the pressure on either side of the eardrum. The *epitympanic recess* connects with air cells in the mastoid process via the *mastoid antrum*.

The Inner Ear (*Fig. 29a and c*)

The inner ear has a double function, being concerned with balance as well as with hearing. The *bony labyrinth*, a tortuous system of tunnels in the petrous temporal bone, encloses the *membranous labyrinth*, which is filled with *endolymph* and surrounded by *perilymph*.

Hearing

The *cochlea* (*Fig. 29c and d*) is the part of the inner ear concerned with hearing. It contains the *cochlear duct*, part of the membranous labyrinth connected to the *saccule*. It is 3 cm long, spiralling for 2¾ turns round a hollow bony pillar, the *modiolus*. The *vestibular membrane* forms the roof of the cochlear duct and the *basilar membrane* its floor. The channel lying above the cochlear duct is the *scala vestibuli*, which is separated from the middle ear by the oval window. Below the cochlear duct lies the *scala tympani*, with the *round window* at its lateral end. The two scalae are filled with perilymph and are continuous with each other around the tip of the cochlear duct. Within the duct is the spiral *organ of Corti*, whence cochlear (auditory VIII) nerve fibres pass to the brain. Movements of the stapes within the oval window produce pressure changes in the perilymph, which result in movement of the basilar membrane. This stimulates the specialized cells of the organ of Corti, the auditory nerve cells fire and hearing information is carried to the brain.

Balance (Fig. 29c)

The parts of the inner ear concerned with balance are the *vestibule* and the *bony semicircular canals*, which form part of the bony labyrinth. The corresponding parts of the membranous labyrinth are the membranous *semicircular canals*, the *utricle* and the *saccule*. The three semicircular canals are set at right angles to each other, and can thus register movement in any direction in space. They have five openings (one is shared) into the utricle, which is connected to the saccule. The ampullae of the semicircular canals contain hair cells, which are bent by the viscous endolymph around them whenever the head is moved.

The utricle and saccule each have a specialized area, the macula, which has hair cells with attached otoliths to weight them. These hair cells are surrounded by nerve endings which fire when the hair cells are stimulated. Information about the position and movement of the head in space is carried from these structures to the brain by the vestibular division of the VIII nerve (*Fig. 29a*).

The Cardiovascular or Circulatory System
(*Figs 30 – 38*)

The heart and blood vessels form a transport system for oxygen, nutrients, waste products, hormones and antibodies within the body.

THE HEART (*Fig. 30b*)

The heart is about the size of its owner's clenched fist. It lies in the centre of the thoracic cavity, behind the sternum and in front of the descending aorta and the oesophagus. Its base points upwards, backwards and to the right and its apex downwards, forwards and to the left.

The heart is enclosed by the *pericardium* (*Fig. 30c*). The outer fibrous pericardium forms a bag attached to the diaphragm and to the outer coats of the great vessels. Within it is the serous pericardium, a sac invaginated by the heart like a fist punched into a soft balloon. Its visceral layer, the *epicardium*, covers the heart and great vessels, and its outer parietal layer lines the fibrous pericardium. Between these two layers is a film of fluid that allows free movement of the heart within the pericardial sac. The heart itself consists of cardiac muscle (*myocardium*), which is thicker on the left side than on the right. The heart cavities are lined by a smooth endothelial layer, the *endocardium*.

The heart has four chambers (*Fig. 30a*), right and left *atria* above and right and left *ventricles* below. The *septum* separates the right and left sides of the heart. The *tricuspid valve* (*Fig. 32a*) (3 leaflets) separates the right atrium from the right ventricle. The *mitral valve* (2 leaflets) separates the left atrium and ventricle. The *chordae tendineae* and *papillary muscles* tether the edges of the valves to the ventricular walls and prevent the valves bursting back into the atria when the ventricles contract in *systole*; relaxation is called *diastole*.

The *sinuatrial (SA) node (Fig. 30c)* is the natural pacemaker of the heart and initiates the heart beat. From it the stimuli for cardiac contraction radiate outwards and downwards through both atria and are picked up by the *atrioventricular (AV) node* in the interatrial septum just above the tricuspid valve. From there impulses travel downwards into the ventricles along the *bundle of His (Purkinje fibres)*, which branches again and again to transmit the impulses to the cardiac muscle. Both atria contract together (atrial systole), after which both ventricles contract simultaneously (ventricular systole), pumping blood into the great vessels. This is followed by diastole, when the relaxed heart fills with blood again.

BLOOD VESSELS (*Fig. 31b*)

Arteries and Arterioles

Arteries carry blood *away* from the heart. They have smaller branches called *arterioles*. All except the pulmonary arteries carry oxygenated blood. This blood is at higher pressure than that on the venous side of the circulation. The arteries have thick walls, in three layers or coats (*Fig. 31b*):

(1) tunica intima, the inner layer of smooth endothelium;
(2) tunica media, the middle layer, which is the thickest, consisting of smooth muscle and elastic tissue;
(3) tunica adventitia, the outer layer, which consists of a coat of collagen fibres which merge with the connective tissue surrounding the vessels.

Arteries lead into arterioles. As they branch to form smaller and smaller vessels, the connective tissue and muscular coats are gradually lost and the remaining branching tubes of endothelium form the *capillary bed*. The capillaries have thin, permeable walls through which the tissues obtain oxygen and nutrients and pass back carbon dioxide and waste products.

Veins and Venules

The capillaries join up to form small veins called *venules*, which in turn unite to form *veins*. Veins carry blood *towards* the heart. Except for that in the pulmonary veins, the blood they carry is dark and deoxygenated.

Veins also have three-layered walls, proportionately thinner than arterial walls. The intima consists of endothelium. The media has little elastic tissue or muscle, but a great deal of collagen. The adventitia also contains muscle and elastic fibres. The walls, though thin, are thus very strong.

Veins over about 2 mm in diameter have *valves* shaped like a pocket with the free edge towards the heart (*Fig. 31b*). When blood flows towards the heart, the valve 'pockets' are empty and lie flat against the vein walls. However, if the blood starts to flow backwards the pockets fill up and block the lumen of the vein, effectively preventing further backward flow.

The veins converge to become fewer but larger, until eventually the *superior* and *inferior venae cavae* are formed, draining into the right atrium (*Fig. 30a*).

Vasomotor nerves constrict and relax the muscle in the blood vessel walls.

Veins and large arteries also have their own blood vessels (*vasa vasorum*) and lymphatics running in the adventitia.

THE CIRCULATION OF THE BLOOD (*Figs 30a and 31a*)

Deoxygenated blood leaves the right ventricle to be carried by the pulmonary trunk and arteries to the lungs, where it eventually reaches the capillaries in the alveolar walls. Here the blood takes up *oxygen* (O_2) from the alveolar air in exchange for *carbon dioxide* (CO_2). *Oxygenated* blood then passes back along the pulmonary veins to the left atrium and having reached the left ventricle, is pumped out via the aorta and arteries to supply all parts of the body. Having given up its oxygen and picked up carbon dioxide, the deoxygenated blood returns via the veins to the right atrium and ventricle, and is again pumped to the lungs for the cycle to be repeated.

The part of the cycle from the right ventricle through the lungs to the left atrium is called the *pulmonary circulation*, whereas the *systemic circulation* supplies blood to the rest of the body via the *aorta* and its branches.

The Aorta (*Figs 30 and 33*)

The ascending aorta starts from the base of the left ventricle and passes up behind the sternum (*Fig. 33*).

The *right* and *left coronary arteries* (*Fig. 30*) branch from it to supply the muscle of the heart itself.

Arterial Supply to the Upper Limbs

The *arch* of the aorta gives off the arteries which supply the head, neck and upper limbs:

(1) the *brachiocephalic trunk* (innominate artery), which bifurcates to form
 (a) the right *common carotid artery* to the right side of the head and neck,
 (b) the right *subclavian artery* to the right upper limb;
(2) the left *common carotid artery* to the left side of the head and neck;

(3) the left *subclavian artery* to the left upper limb.

Arterial Supply to the Trunk

The *descending* aorta has two parts: thoracic and abdominal.

Thoracic

The thoracic aorta and its branches supply the chest through:

(1) *bronchial* arteries to the lungs and bronchi;
(2) *pericardial* branches to the pericardium;
(3) *oesophageal* branches to the oesophagus;
(4) *posterior intercostal* arteries (9 pairs), supplying the muscles in the lower 9 intercostal spaces;
(5) *phrenic* arteries to the upper surface of the diaphragm.

Abdominal

The abdominal aorta enters the abdomen behind the diaphragm and descends in front of the vertebral column, dividing into the two common iliac arteries.

Branches of the abdominal aorta (*Fig. 33*) are:

(1) *phrenic* arteries to the lower surface of the diaphragm;
(2) the *coeliac* artery (axis) (*Fig. 34a*), with its own branches:
  (*a*) the *common hepatic* artery to the liver and gallbladder. It also gives off the *right gastric* artery and the *gastroduodenal* artery;
  (*b*) the *left gastric* artery to the stomach;
  (*c*) the *splenic* artery to the spleen and pancreas;
(3) *suprarenal* arteries (2) to the suprarenal glands;
(4) the *superior mesenteric* artery (*Fig. 34b*) supplying the midgut;
(5) the 2 *renal* arteries to the kidneys;
(6) *gonadal* arteries (2): *testicular* (male); *ovarian* (female);
(7) the *inferior mesenteric* artery (*Fig. 34c*) supplying the hindgut;

(8) *lumbar* arteries (5 pairs) to the lumbar muscles and abdominal wall;
(9) the *median sacral* artery to the sacrum.

The *common iliac* arteries run downwards and laterally, supplying small branches to the structures around them. They end by dividing into internal and external iliac arteries at the level of the disc between the last lumbar vertebra and the sacrum.

The *internal iliac* arteries (*Fig. 34d*) supply the pelvic viscera as follows:

(1) *vesical* arteries: bladder;
(2) *middle rectal* (*haemorrhoidal*) arteries: rectum;
(3) *uterine* arteries (in the female): uterus and upper part of the vagina;
(4) *vaginal* arteries: the rest of the vagina;
(5) *internal pudendal* arteries: lower part of the anal canal and the external genitalia in both sexes;
(6) *obturator* arteries: pelvic wall and organs;
(7) *superior* and *inferior gluteal* arteries: buttocks and back of thigh;
(8) *lateral sacral* arteries: sacral vertebrae and their overlying tissue.

Arterial Supply to the Lower Limbs

The *external iliac* arteries run downwards and outwards to enter the thighs as the *femoral* arteries, whose branches supply the lower limbs.

Arterial Supply to the Head and Neck (*Figs 33 and 35a*)

The *left common carotid* (from the aorta) and the *right common carotid* (from the brachiocephalic trunk) ascend on either side of the neck and divide into *external* and *internal carotid* arteries.
Branches of the external carotid are:

(1) *superior thyroid* arteries to the thyroid gland;
(2) *lingual* arteries to the tongue;
(3) *facial* arteries running upwards from behind the angle of the jaw to supply the inside of the mouth cavity, the submandibular gland and tonsil and the adjacent tissues of the upper neck and lower face;
(4) *occipital* arteries supplying the ears and mastoid, part of the scalp, the pharynx, soft palate and tonsils.

In front of the ears the external carotid arteries divide to form:

(5) *superficial temporal* arteries to the parotid gland, temporomandibular joint and part of the face and scalp;
(6) *maxillary arteries* supplying the jaws, muscles of mastication, the palate, teeth and nose;
(7) *middle* and *small meningeal* arteries which supply the dura mater around the brain.

Arterial supply to the Brain (*Figs 35a and b*)

This happens via the left and right internal carotid arteries and the left and right vertebral arteries.

The *vertebral* arteries (*Fig. 35b*) arise from the subclavians and, running upwards through the foramina in the transverse processes of the upper six cervical vertebrae, enter the skull through the foramen magnum. At the pons, they join to form the *basilar* artery. This gives off branches to the cerebellum, pons and medulla, and divides into two *posterior cerebral* arteries.

The *internal carotid* arteries supply the cerebral hemispheres, eyes, forehead and nose. They run upwards from the bifurcation of the common carotid to the base of the skull, and then forward through the carotid sinuses, giving off the *ophthalmic* arteries to the eyes as they leave, and end by dividing into *anterior* and *middle cerebral* arteries.

An arterial anastomosis, the *circle of Willis*, joins the vertebral and carotid arteries at the base of the brain around the optic chiasma and pituitary stalk. This circle enables the circulation to be maintained if one of the main feeding vessels is blocked, provided the circle itself is not obstructed by vascular disease.

Venous Return (*Fig. 36*)

The superior and inferior venae cavae return deoxygenated blood to the right atrium from the upper and lower half of the body respectively, while the *coronary sinus* drains the heart muscle itself.

Veins of the Face and Scalp (Fig. 36)

These run alongside the arteries and have the same names. They drain either into the *external jugular* vein and thence to the *subclavian* vein, or into tributaries of the *internal jugular* vein.

Venous Drainage of the Brain (*Fig. 37*)

The brain veins empty into the *venous sinuses* of the dura mater, which lie between its layers and are lined by endothelium continuous with the lining of the veins. Their blood drains into the *sigmoid sinuses*, which leave the skull through the jugular foramina and become the *internal jugular* veins.

The *cavernous sinuses* (*Fig. 24c*) drain blood from the eyes, face and part of the brain, and the two sides communicate across the midline. They drain through the *superior* and *inferior petrosal sinuses* to the internal jugular veins.

Venous Return from the Upper Limbs

All the venous blood returning from the upper limbs flows into the *subclavian* vein, which joins with the *internal jugular* vein to form the *brachiocephalic trunk* (innominate vein). The right and left brachiocephalic trunks unite to form the *superior vena cava*.

Venous Return from the Lower Limbs

The *deep* veins accompany the arteries and are similarly named. The *superficial* veins are the *great* and *small saphenous* veins and their tributaries. All the blood from superficial and deep veins ends in the *femoral* vein which continues upwards behind the inguinal ligament to become the *external iliac* vein. The *internal iliac* vein receives blood from the pelvic organs, gluteal region and genitalia and joins with the external iliac to form the *common iliac* vein.

The two common iliac veins run obliquely upwards and join to form the *inferior vena cava*. This ascends in front of the vertebral column receiving blood from the spine, gonads, kidneys, suprarenals and liver, and runs behind the liver to perforate the diaphragm and pericardium and end in the right atrium.

The Portal System of Veins (*Fig. 38*)

This is the means by which food substances absorbed from the gut are conveyed to the liver for processing. The *portal* vein enters the organ at the *porta hepatis*. The veins comprising it and the organs they drain are:

(1) *superior mesenteric* vein (small intestine and part of the large intestine);
(2) *inferior mesenteric* vein (rest of large intestine);
(3) *gastric* veins (stomach);
(4) *splenic* vein (spleen and pancreas).

The inferior mesenteric drains into the splenic vein which then joins the superior mesenteric to form the portal vein, which runs upwards to enter the liver.

The Lymphatic System
(Fig. 39)

This consists of:

(1) *lymph vessels (lymphatics)*;
(2) *lymphoid tissue*, found in lymph nodes, spleen, thymus, small intestine, appendix and bone marrow. The tonsils and adenoids are specialized masses of lymphoid tissue.

The lymph vessels collect tissue fluid exuded from the capillaries and return it to the general circulation, while the lymph nodes filter the fluid, removing dust, infective matter and malignant cells. The nodes also help produce lymphocytes and antibodies.

LYMPHATICS (*Fig. 39a*)

These are thin-walled tubes, usually running alongside blood vessels. Their blind ends are bathed in tissue fluid and equipped with valves; when full they appear beaded. Flow within them is achieved by the squeezing action of the surrounding muscles.

Lymphatics occur everywhere in the body **except** in the

(a) cornea
(b) central nervous system
(c) striated muscle
(c) bone marrow
(e) joint cartilage
(f) hair and nails.

The small lymphatics join to form larger vessels, which have lymph nodes along their length. Eventually all the lymph is collected into two large channels, the thoracic duct and the right lymphatic duct, which return it to the venous system.

The *thoracic duct* arises as a dilatation, the *cisterna chyli*, opposite the 2nd lumbar vertebra (L2), and then ascends through the abdomen and chest to open into the venous system where the left internal jugular vein meets the left subclavian vein. It carries lymph from both lower limbs, the pelvis, abdomen, neck and chest, the left arm and the left side of the head.

The *right lymphatic duct* drains the right side of the head, neck, chest and right arm, and opens into the bloodstream at the junction of the right internal jugular and right subclavian veins.

LYMPH NODES (*Fig. 39a, b and c*)

The main lymph nodes and the areas draining to them are:

(*a*) occipital: back of the scalp
(*b*) retroauricular: area round the ear
(*c*) submental and submandibular: face and the floor of the mouth
(*d*) superficial cervical lymph nodes: external ear and neck
(*e*) deep cervical nodes: tongue, the lower pharynx
(*f*) axillary nodes: upper limbs and breast
(*g*) epitrochlear nodes: hands and forearms
(*h*) sternal and intercostal nodes: chest wall and upper abdominal wall
(*i*) mediastinal nodes: thoracic structures
(*j*) tracheobronchial nodes: lungs
(*k*) coeliac nodes: stomach, spleen, pancreas and liver
(*l*) mesenteric nodes: intestines
(*m*) lumbar nodes: kidneys, adrenals and ovaries or testes
(*n*) internal iliac nodes: pelvic viscera
(*o*) external iliac nodes: groins

(p) superficial inguinal nodes: buttocks, anus, perineum and external genitalia. They receive nearly all the superficial lymph from the lower limbs
(q) deep inguinal nodes: deep tissues of the lower limbs
(r) popliteal nodes: legs below the knees.

Collections of lymphoid tissue (*Peyer's patches*) are found in the walls of the alimentary canal and appendix.

The Spleen

The spleen (*Fig. 14*) is a dark-purple, bean-shaped organ lying high up in the left abdomen behind the stomach and against the diaphragm. It is usually about 12.5 cm long, 7.5 cm broad and 4 cm thick. It has a fibrous capsule from which *trabeculae* (fibrous bundles) pass into the *pulp*. The splenic artery and vein enter and leave it at its hilus. It produces *lymphocytes* and *phagocytes* and destroys old red blood cells.

The Thymus

The thymus, which contains lymphoid tissue, has already been described (*see p. 82 and Fig. 19*).

The Skin and its Appendages
(*Fig. 40a*)

The skin, which covers the whole of the body, is protective, elastic, waterproof, sensitive and helps in temperature regulation. It consists of an outer *epidermis*, sharply marked off from the underlying *dermis*. This blends with the loose connective tissue beneath it, joining the skin to the underlying fascia. It contains variable amounts of fat, which conserve heat and act as a reserve energy store.

THE EPIDERMIS AND THE DERMIS (*Fig. 40a*)

The epidermis varies in thickness over the body; it is thickest over the palms of the hands and soles of the feet, and thinnest on the flexor surfaces of the limbs. Its uppermost layer, the horny *stratum corneum*, consists of layers of flat, clear scales which are constantly shed. The lost cells are continually replaced by cells pushed up from below.

Where the epidermis and dermis meet, there are numbers of *melanocytes* (black cells) with branching processes that run among the cells of the *stratum spinosum* and may produce their *melanin*.

The dermis or *corium* is dense and closely adherent to the epidermis above, becoming looser below. Its thickness is variable and it has ridges called *papillae*, which are most numerous and regular over the fingertips, palms and soles. The *epidermal ridges* that overlie them produce the individual's characteristic fingerprint, palm print and sole print. Beneath the papillae are *collagen bundles* which are particularly strong in parallel to the lines of tension of the skin and the skin creases. Smooth muscle fibres are present in the dermis in the penis, scrotum and nipples and in relation to the hair follicles and sweat glands.

Arterial supply

This is from the subcutaneous tissues to capillary networks within the dermal papillae.

Venous drainage

This is via plexuses into subcutaneous veins.

There are no blood vessels in the epidermis; it is nourished by tissue fluid from the dermis.

Nerve supply

The skin has a very rich afferent nerve supply and contains specialized nerve endings for detecting touch, stretch, pain, pressure and temperature. In addition, efferent *vasomotor* nerves run to the blood vessels, *secretomotor* fibres to the sweat glands and *pilomotor* ones to the hairs.

The Nails

Nails consist of modified epidermis. Each has a free edge, which grows beyond the end of the finger or toe, a body which lies against the nail bed and is pink except for the white *lunula* (*half-moon*), and a root, which is covered by skin. At the base of the nail the stratum corneum grows over it for a short distance as the *eponychium* (*cuticle*). The nail grows from the lunula and is pushed forwards from it. The nail bed is richly innervated.

Sweat Glands (*Fig. 40a*)

Each consists of a tiny tube rolled into a ball, lying deep in the dermis. The duct runs up to reach the epidermis and finally opens on the surface at a sweat pore. The glands extract water, salts and waste products from surrounding dermal capillaries and discharge the product on the skin surface as *sweat*. They are supplied by non-myelinated nerve fibres and are embraced by *myoepithelial cells*, which can squeeze them.

Some parts of the body have specialized sweat glands. These are:

(1) *ceruminous* glands in the external auditory meatus which produce the fatty earwax;
(2) *apocrine* glands in the axilla and around the anus, whose lining cells disintegrate to produce their secretion, which has a characteristic smell;
(3) *ciliary* glands in the eyelid.

Hairs (*Fig. 40a*)

Hairs are developed from the epidermis, and consist of a *shaft* of *keratin* projecting above the skin surface and the hair *root* enclosed by the skin. Deep in the dermis the root expands into a hair *bulb*, which is moulded over a papilla containing nerves and a capillary loop. The hair *follicle* encloses the root, and *sebaceous glands* open into its upper part. The connective tissue sheath of the follicle is connected to the dermal papilla by a band of unstriped muscle, the *arrector pili*. The hair is raised upright and the skin is pulled into 'goosepimples' by this muscle.

Sebaceous glands are small sac-like glands usually associated with hairs, one or more glands opening into each hair follicle. Secretion occurs by fatty disintegration of cells to produce the greasy *sebum*, which passes into the hair follicles and thence to the skin surface.

THE BREASTS (Mammary Glands) (*Fig. 40b*)

These are hemispherical and lie within the superficial fascia on the front and sides of the chest, with an axillary tail extending upwards on each side. The bulk of the breast consists of fat, giving it a rounded contour. Fibrous tissue septa divide the glandular tissue, which consists of 15–20 lobes, arranged like the spokes of a wheel. Each lobe has lobules of milk-producing *alveoli*, draining into *lactiferous ducts* which converge on the *areola*, the pigmented area surrounding the *nipple*. The areola is studded with lubricating alveolar glands. Circular plain muscle fibres around the nipple can erect it.

The breast is rudimentary at birth, and in the male develops no further. The female breast shows a growth spurt at puberty, and becomes fully developed during pregnancy in preparation for lactation.

Arterial supply

From branches of the internal thoracic and intercostal arteries which pierce the intercostal spaces, and from the thoracic branches of the axillary arteries.

Venous drainage

To the axillary and internal thoracic veins.

Lymphatic drainage

Mainly to the axillary nodes, but some vessels communicate with the parasternal lymph nodes and with the lymph vessels of the opposite breast.

Nerve supply

From branches of 4th, 5th and 6th thoracic nerves (T4, T5 and T6) which carry both sympathetic and somatic fibres.